T0281530

Mathematical Physics for Nuclear Experiments

Mathematical Physics for Nuclear Experiments

Andrew E. Ekpenyong

CRC Press

Taylor & Francis Group

Boca Raton London New York

CRC Press is an imprint of the
Taylor & Francis Group, an **informa** business

First edition published 2022
by CRC Press
6000 Broken Sound Parkway NW, Suite 300, Boca Raton, FL 33487-2742

and by CRC Press
2 Park Square, Milton Park, Abingdon, Oxon, OX14 4RN

Library of Congress Cataloging-in-Publication Data

Names: Ekpenyong, Andrew E., author.
Title: Mathematical physics for nuclear experiments / Andrew E. Ekpenyong.
Description: First edition. | Boca Raton : CRC Press, 2022. | Includes
bibliographical references and index.
Identifiers: LCCN 2021020492 | ISBN 9780367768522 (hardback) | ISBN
9781032104997 (paperback) | ISBN 9781003215622 (ebook)
Subjects: LCSH: Mathematical physics. | Nuclear physics--Experiments.
Classification: LCC QC20 .E37 2022 | DDC 530.15--dc23
LC record available at https://lccn.loc.gov/2021020492

ISBN: 978-0-367-76852-2 (hbk)
ISBN: 978-1-032-10499-7 (pbk)
ISBN: 978-1-003-21562-2 (ebk)

DOI: 10.1201/9781003215622

Typeset in Latin Modern font
by KnowledgeWorks Global Ltd.

eResources are available for this title at https://www.crcpress.com/9780367768522

Dedication

I dedicate this work to Creighton University Physics Professors David Sidebottom, Gintaras Duda, Jack Gabel, Janet Seger, Michael Cherney, Michael Nichols, Robert Kennedy (RIP) and Sam Cipolla, who were my teachers and now have me as a colleague.

Contents

x ■ Contents

Preface to the first edition

This book presents some mathematical derivations of the equations used in describing and analyzing results of typical nuclear physics experiments for advanced undergraduates and early graduates. The physical intuitions behind the derivations of these equations are retained. The intended audiences for this text include students taking Nuclear Instruments and Methods (NIM), Medical Physics and Experimental Particle Physics courses. It was borne out of the author's lecture materials used in elucidating standard experiments in a NIM course for advanced undergraduates and graduate students, which the author still teaches at Creighton University, Omaha, NE, USA. Figures and tables showing results from the author's experiments and those of his students are used in demonstrating experimental outcomes. Instead of merely using the results and citing texts, equations such as the Bohr's classical formula and Bethe-Bloch's quantum mechanical formula for energy loss of heavy charged particles, Poisson, Gaussian and Maxwellian distributions for radioactive decay, Fermi function for beta spectrum analysis, Tamm-Frank formula for energy of Cherenkov radiation, Klein-Nishina quantum electrodynamics formula for Compton scattering, and so on, are presented with the mathematical bases of their derivation and with their physical utility. This strategy provides greater connection between theoretical and experimental nuclear physics. Furthermore, connections are made between well-established results and ongoing research. For instance, we show in this book that the well-established Bateman equations have recently been solved using matrix exponential methods which are more computer-friendly, making further work not only easier but attractive. It is hoped that such connections and attractions will help prepare scientists who will bring progress to both experimental and theoretical nuclear physics research as well as novel applications therefrom, especially in medical physics.

About the Author

Dr Andrew E. Ekpenyong, summarizing some equations

Andrew Ekpenyong earned his PhD in Physics from the University of Cambridge, UK, in 2012. He obtained a Master of Science degree in Physics in 2007 from Creighton University, Omaha, Nebraska, USA, where he is currently a professor. Dr Ekpenyong teaches Quantum Mechanics, Nuclear Instruments and Methods, Dosimetry and Radiation Protection, Physics of Radiotherapy, Biophysics as well as General Physics. He has supervised research and taught both graduates and undergraduates at Technical University, Dresden, Germany and Creighton University, Omaha, USA. Dr Ekpenyong has authored/co-authored several articles in peer-reviewed journals in physics and medical physics. His research and ensuing discoveries have ranged from the physics of cancer (a new frontier bordering on the mechanical properties of cancer cells and their role in cancer disease and metastasis) to the physics of radiation therapy. In jocular fashion, his research is in the surrounding of the cell nucleus and the atomic nucleus. A bit more seriously, this is part of the field of biomedical physics.

The author hereby designates that all royalties from this work be donated to the 100% free hospitals which he founded in rural areas of south-south Nigeria to help the less privileged: Joseph Ukpo Hospitals and Research Institutes with url: https://www.juhri.org/.

Acknowledgements

I am grateful to all my students who took the courses Nuclear Instruments and Methods (NIM), Physics of Radiation Therapy and Dosimetry/Radiation Protection. I thank Prof Sam Cipolla who taught the course NIM for over 40 years and handed over the baton to me in 2015, while still available as Emeritus Professor, helping with setup and even co-teaching some aspects of the course. We are still using the Lab Manual he prepared for the details of the experiments in the course. I sincerely thank the staff of Institute of Physics and those of the Publisher for their professionalism and attention to detail. Special gratitude goes to Jeanine Burke, the Consulting Editor of IOP Concise Physics e-book program, and to Dr Kirsten Barr, the Physics Editor of CRC Press/Taylor and Francis Group. Their consistent prodding ensured my completion of this work. I thank all my colleagues in the Creighton University Physics Department. They have provided for me a home away from home.

Symbols

Table 0.1 Symbols, Meanings, Values and Units

Symbol	Meaning	Values/Units/Remarks
α	alpha particle, Helium nucleus,	heavy charged particle, $+2e$
α	fine structure constant	$\frac{1}{137.03599976(50)}$
β	beta particle,	light charged particle, $\pm e$
β	relative speed v/c	dimensionless
γ	gamma ray, photon,	uncharged
γ	Lorentz factor $\frac{1}{\sqrt{(1-\beta^2)}}$	dimensionless
δ	density effect correction to energy loss	dimensionless
ε	permittivity	F/m
λ	decay constant, wavelength	s^{-1}, m
ρ	density	kg/m^3, g/cm^3
c	speed of light	299792458 m/s
T	kinetic energy	J, MeV
$m_e c^2$	electron mass$\times c^2$	0.511 MeV
r_e	classical electron radius $\frac{e^2}{4\pi\varepsilon_0 m_e c^2}$	2.8179 fm
v_0	Bohr velocity $\frac{e^2}{\hbar}$	25 keV/u
N_A	Avogadro's number	6.02214×10^{23} mol^{-1}
A_1	atomic mass of incident particle	g mol^{-1}
A_2	atomic mass of target atom	g mol^{-1}
M_1	mass of incident particle	MeV/c^2 or u
M_2	mass of target atom	MeV/c^2 or u
N_ε	electron density	units of $(r_e)^{-3}$
E_c	critical energy	MeV
X_0	radiation length	g cm^2

Table 0.2 Symbols, Meanings, Values and Units

Symbol	Meaning	Values/Units/Remarks
I	mean excitation energy (ionization potential)	eV
K/A	stopping prefactor $4\pi N_A r_e^2 m_e c^2/A$	0.307075 cm^2 g^{-1}
S	stopping power	
v	projectile particle velocity	m/s
w_j	weight fraction of jth element in a compound/mixture	dimensionless
Z_1	projectile particle charge number in units of electron charge e	e
Z_2	target material atomic number	
λ_c	Compton wavelength $\lambda_c = \frac{h}{mc}$	m
λ_c	Compton wavelength of electron $\lambda_c = \frac{h}{m_e c}$	2.426310×10^{-12} m
$\lambdabar$	reduced Compton wavelength $\frac{\lambda_c}{2\pi}$	
$\lambdabar_e$	reduced Compton wavelength for electron $\frac{\lambda_c}{2\pi}$	386 fm
a_0	Bohr radius $\frac{4\pi\varepsilon_0 \hbar^2}{m_e e^2}$	5.291772×10^{-11} m
CSDA range	continuous slowing down approximation range	g/cm^2
$\frac{S}{\rho} = \frac{dE}{\rho dx}$	Mass stopping power	MeVg^{-1}cm^2
$\frac{W_{air}}{e}$	ionization constant	33.97 eV/ion pair in air
amu or u	unified atomic mass unit	931.494 MeV/amu = 1 g/mole
amu or u	atomic mass unit	1.660539×10^{-27} kg
amu or u	energy mass conversion	931.494 MeV/c^2
m_e	mass of the electron/positron	0.511 MeV/c^2 = 9.11×10^{-31} kg
q_e or e	charge of the electron	1.60218×10^{-19} C
m_p	mass of the proton	938 MeV/c^2 = 1.6726×10^{-27} kg
m_n	mass of the neutron	939.57 MeV/c^2 = 1.6749×10^{-27} kg
m_α	mass of the alpha particle	3727.33 MeV/c^2 = 6.645×10^{-27} kg

CHAPTER 1

Radioactivity and Decay Law

Alan Turing's words that science is a differential equation seem quite true in the physics of radioactivity. This chapter begins with a differential equation called the radioactive decay law and ends with equations for specific decay processes in the unstable nucleus including alpha decay, beta decay and gamma decay. Throughout the chapter, we give mathematical derivations, equations and example calculations which enable us to make sense of (theory) and make use of (applications) the results of nuclear experiments.

1.1 THE RADIOACTIVE DECAY LAW

The radioactive decay law simply states that the activity of a radioactive sample decays exponentially in time. This law implies that the rate at which a radioactive sample decays (activity) is proportional to the remaining number of atoms. Derived experimentally by Ernest Rutherford (1871–1937) and Frederick Soddy (1877–1956), it can also be derived from Quantum Mechanics in terms of a transition probability per unit time, λ, which is characteristic of the nuclear species. Since λ is a constant, independent of time, it is aptly termed decay constant. Thus, in a sample of $N(t)$ radioactive nuclei, Eq. 1.1a gives the mean number of nuclei dN, decaying in a time dt:

$$dN = -\lambda N dt. \qquad (1.1a)$$

$$\frac{dN}{dt} = -\lambda N. \qquad (1.1b)$$

$$\frac{dN}{N} = -\lambda dt. \tag{1.1c}$$

Eq. 1.1b states the decay law as a differential equation, making it obvious as a separable equation which can therefore be written in the form given by Eq. 1.1c. Integrating both sides of Eq. 1.1c, we get

$$\int_{N_0}^{N} \frac{dN}{N} = -\lambda \int_{t_0}^{t} dt \tag{1.2}$$

which reduces to

$$\ln N - \ln N_0 = -\lambda (t - t_0).$$

Now, N_0 is $N(t)$ at t_0, an arbitrary initial time. We then set $t_0 = 0$ to obtain,

$$\ln \left(\frac{N}{N_0} \right) = -\lambda t.$$

Finally, taking the exponential of both sides and rearranging, we obtain $N(t)$ in Eq. 1.3:

$$\boxed{N(t) = N_0 e^{-\lambda t}.} \tag{1.3}$$

The *decay law*, expressed in Eq. 1.3, gives the number of "survivors", $N(t)$ of radioactive decay. That is, the initial number of nuclei N_0 drops to $N(t)$ at time t. For emphasis, the *fraction of survivors* (those that have not decayed) is:

$$\frac{N(t)}{N_0} = e^{-\lambda t}. \tag{1.4}$$

Obviously, the fraction of nuclei that has decayed in time t is given by Eq. 1.5:

$$\frac{N_0 - N(t)}{N_0} = 1 - e^{-\lambda t}. \tag{1.5}$$

From the *decay law*, the rate of decay or *activity A* of a radioactive sample is thus determined by its decay constant, λ. From Eq. 1.1b, $A = \lambda N$, thereby following the same time dependence as $N(t)$ in Eq. 1.3. It turns out that the *mean lifetime* or average lifetime or just mean life of a radioactive nucleus, τ, is just $1/\lambda$. This can be shown mathematically as follows. From Eq. 1.1b, the number of nuclei decaying in the interval between t and $t + dt$ is

$$dN = \lambda N(t) dt = \lambda N_0 e^{-\lambda t} dt.$$

Thus, applying the standard method for finding the mean of a continuous variable, as well as integration by parts, we have:

$$\tau = \frac{\int t\,dN}{\int dN} = \frac{\int_0^\infty t e^{-\lambda t}\,dt}{\int_0^\infty e^{-\lambda t}\,dt} = \frac{\frac{1}{\lambda^2}}{\frac{1}{\lambda}} = \frac{1}{\lambda}. \tag{1.6}$$

Intuitively, Eq. 1.6 shows that the mean lifetime τ of a nuclide (nucleus plus electrons) is the sum of the life times of a given number of nuclei before they have all disintegrated, divided by the number of nuclei. In practice, the *half-life*, $T_{1/2}$ is very much in use and is defined as the time it takes for the sample to decay to $1/2$ its initial activity. Using Eq. 1.3,

$$\frac{N(t=T_{1/2})}{N_0} = \frac{1}{2} = e^{-\lambda T_{1/2}}. \tag{1.7}$$

This gives,

$$\ln\left(\frac{1}{2}\right) = -\lambda T_{1/2}. \tag{1.8}$$

Solving further,

$$0 - \ln 2 = -\lambda T_{1/2}. \tag{1.9}$$

Hence,

$$T_{1/2} = \frac{\ln 2}{\lambda}. \tag{1.10}$$

Succinctly, the *half-life $T_{1/2}$, mean lifetime τ* and decay constant λ are related by solving for $T_{1/2}$ in Eq. 1.7 and combining with Eq. 1.6:

$$T_{1/2} = \tau \ln 2 = \frac{\ln 2}{\lambda} \approx \frac{0.693}{\lambda} \tag{1.11a}$$

$$\text{or } \tau = \frac{T_{1/2}}{\ln 2} \approx 1.44 T_{1/2}. \tag{1.11b}$$

Thus, $\tau > T_{1/2}$. More explicitly, just as $T_{1/2}$ is the time it takes for the nuclides to decay to $1/2$ the original number, the mean lifetime is the time it takes for the sample to decay to $1/e$ of its initial activity (Eq. 1.12):

$$\frac{N(t=\tau)}{N_0} = e^{-\lambda \tau} = \frac{1}{e} \approx 0.368. \tag{1.12}$$

Example 1.1 *Eq. 1.11a allows us to determine the decay constant λ by measuring the half-life using detectors that record counts, for radionuclides with long half-lives (which means that neither their activity nor their number is changing significantly during the period of measurement). Given that the half-life of Cs-137 is about 30 years, what is it's decay constant?*

Solution

The solution is straightforward.

$$\lambda_{Cs^{137}} = \frac{\ln 2}{(30y)(365d/y)(24h/d)(60min/h)} = 4.4 \times 10^{-8}\frac{1}{min}.$$

So if one waits for a minute, the probability p that a particular radioactive nucleus of Cs-137 will decay is only 4.4×10^{-8} or 1 in 23 million. But this is not a big problem. Nuclei available to decay are very many, of the order of 10^{23}! Note that

$$p = \lambda \Delta t.$$

Example 1.2 *A common unit for measuring the activity $\mathscr{A}$ or decay rate R of a radioisotope is the Curie (Ci) where 1 Curie equals 37 billion disintegrations per second (dps): 1 $Ci = 3.7 \times 10^{10}$ dps. Now, the SI unit of activity is Becquerel (Bq) and 1 $Bq = 1$ dps. If the activity of a typical sample of a radioisotope used in laboratory experiments is 0.5 μCi, what is this activity in Becquerel?*

Solution

$$\mathscr{A} \ Bq = 0.5 \times 10^{-6} \ Ci \times (\frac{3.7 \times 10^{10} \ dps}{1 \ Ci}) = 18500 \ Bq$$

In practice, the *specific activity*, $\mathscr{A}_f$, is sometimes needed and is defined thus:

$$\mathscr{A}_f \equiv \frac{Activity}{mass} = \frac{\lambda N}{NM/N_A} = \frac{\lambda N_A}{M} = \frac{\lambda N_A}{A}, \tag{1.13}$$

where M is the molecular weight of the sample (often in g/mole) and N_A is Avogadro's number, here denoting nuclei/mole and A is the atomic mass number of the nuclide. Thus, the specific activity $\mathscr{A}_f$ has SI unit Bq/kg and old unit Ci/g.

Example 1.3 *Radium-226 has a half-life of 1,600 years. What is the activity $\mathscr{A}$ of 1 g of Radium?*

Solution

Since $\mathscr{A} = \lambda N$, we need to find λ and $N(t)$. From Eq. 1.11a,

$$\lambda = \frac{\ln 2}{T_{1/2}}. \tag{1.14}$$

$$\lambda_{Ra^{226}} = \frac{\ln 2}{(1600y)(365d/y)(24h/d)(60min/h)(60s/min)}$$
$$= 1.3737 \times 10^{-11} s^{-1}.$$

Now to get N atoms, we note that the given 1 g of Ra-226 implies that the atomic mass is 226 g, meaning that there are 226 grams of Ra-226 in a gram-mole (mole) of Ra, and there are $N_A = 6.022 \times 10^{23}$ atoms of any substance in a gram mole of that substance, where N_A is Avogadro's number. Hence, the number of atoms of Ra-226 in 1 g is

$$N = \frac{(6.022 \times 10^{23} \text{atoms/g-mole})}{(226\text{g/g-mole})} = 2.6646 \times 10^{21} \text{atoms/g}.$$

Therefore,

$$\mathscr{A} = \lambda N = (1.3737 \times 10^{-11} s^{-1}) \times 2.6646 \times 10^{21} \text{ dps} = 3.66 \times 10^{10} \text{ dps}.$$

This is interesting. This activity is slightly less than 1 Ci. But historically, 1 Ci was defined as the activity of 1 g of Ra-226. The reason is that the number of atoms or particles in 1 mole of a substance, N_A, was determined more accurately after the definition of the Ci. Hence, the new definition of 1 Ci as 3.7×10^{10} dps exactly.

Exercise 1

To study alpha-decay in a Nuclear Instruments and Methods Laboratory, a Polonium-210 source was used with the label showing an activity of 0.1 μCi on the date it was prepared. A student measured the radioactivity of this source with a Geiger counter and observed 1850 Bq. The student noticed that the source was prepared 69 days before the laboratory experiment. (a) What is the expected activity in Bq and in μCi? (b) Why is the measured activity less than expected? The half-life of Po-210 is 138 days.

Exercise 2

Cobalt-60 decays to Nickel with a half-life of 5.272 years. (a) Find the decay constant for the radioactive disintegration of Co-60. (b) What percentage (or fraction) of a sample of Co-60 will remain after 15 years? (c) How long will it take 98 per cent of the original sample to decay?

Exercise 3

Iodine-125 has a half-life of 60 days. Starting with 100 mCi of Iodine-125, what is the activity after 30 days?

Exercise 4

Some nuclides have more than one mode of decay in which case λ is the sum of separate decay constants for each mode of decay. For instance, Potassium-40 undergoes beta minus with λ_1 and beta plus decay with λ_2 as well as electron capture with λ_3 (see following sections for types of beta decay and electron capture). If

$$\lambda_{total} = \lambda_1 + \lambda_2 + \lambda_3, \tag{1.15}$$

what is the total half-life of Potassium-40?

Exercise 5

After a period of time t, the fraction of the initial number of radioactive atoms remaining is...?

A. λt.   B. $1 - e^{-\lambda t}$.   C. $e^{-\lambda t}$ D. λ.   E. $\lambda e^{-\lambda t}$

Exercise 6

Positron emission tomography (PET) is a widespread method of diagnostic imaging which gives information about the metabolism or functioning of organs in the body. In a PET imaging session, a patient is injected with 20 mCi (740 MBq) of 18F-fluorodeoxyglucose or 18 FDG (that is, fluorodeoxyglucose, FDG, labelled with Fluorine 18, a positron emitting radionuclide). F-18 has $T_{1/2} = 109.7$ minutes. The patient is kept for 45 minutes for the FDG to get distributed within the patient's body. What is the activity of F-18 at the end of the 45 minutes, before the imaging is done.

Exercise 7

Given that the half-life of F-18 is 110 minutes, find the number of atoms of F-18 in 370 MBq of 18F-fluorodeoxyglucose. Hint: Since $A = \lambda N$, find N.

Exercise 8

The introduction of radionuclides into the body for diagnostic or therapeutic reasons constitutes nuclear medicine. The dose delivered by such radionuclides depends on their effective half-life, $T_{1/2eff} = \frac{ln2}{\lambda_{eff}}$. The effective decay constant is a combination of radioactive decay and biologic excretion given by

$$\lambda_{eff} = \lambda + \lambda_{bio}.$$

Now, beginning from an initial activity A, it can be shown [93] that the accumulated dose is proportional to the number of atoms that decay in the body such that

$$\text{Dose} \propto \int_0^{t \to \infty} A e^{-\lambda_{eff}t} = \lim_{t \to \infty} \frac{A}{\lambda_{eff}}(1 - e^{-\lambda_{eff}t}). \qquad (1.16)$$

This implies that relative uncertainty of the effective half-life will propagate linearly with the dose. Since the biological half-life cannot be determined as precisely as the physical half-life, argue that for many long-lived nuclides such as Na-22, Co-60, Cs-137, the much shorter biological half-life (10 to 70 days) dominates.

Exercise 9

A radionuclide used in nuclear medicine has a biological half-life that is equal to its physical half-life of 2 hours. What is its effective half-life?. Hint: Since $T_{1/2eff} = \frac{ln2}{\lambda_{eff}}$ and

$$\lambda_{eff} = \lambda + \lambda_{bio},$$

we have that $\frac{1}{T_{1/2eff}} = \frac{1}{T_{phys}} + \frac{1}{T_{bio}}$.

1.2 RADIOACTIVE DECAY CHAIN

The mathematical model that describes activities and products in a decay chain as a function of time is given by the *Bateman equation*. It is based on the decay rates and initial abundances. Ernest Rutherford formulated the essential phenomenological elements of the model in 1905. Harry Bateman provided the analytical solution in 1910 [6]. Let's develop the Mathematics from instances of the physical phenomenon of decay chains. Consider the case of radium which decays into radon which in turn decays to polonium. Starting with pure radium as N_0 at $t = t_0$, we may wish to find the amount of radium (N_1), radon (N_2) and/or polonium (N_3) left at any t. Using Eq. 1.1b and Eq. 1.3 we have for radium,

$$\frac{dN_1}{dt} = -\lambda_1 N_1. \tag{1.17a}$$

$$N_1 = N_0 e^{-\lambda_1 t}. \tag{1.17b}$$

Obviously, the rate of radium decay, $\lambda_1 N_1$ or $\lambda_1 N_0 e^{-\lambda_1 t}$ is the rate of radon creation. But the radon itself is decaying at a rate $\lambda_2 N_2$. Subtracting the decay rate from the creation rate, we have for radon:

$$\frac{dN_2}{dt} = \lambda_1 N_1 - \lambda_2 N_2. \tag{1.18}$$

Eq. 1.18 can be rearranged to obtain:

$$\frac{dN_2}{dt} + \lambda_2 N_2 = \lambda_1 N_1 = \lambda_1 N_0 e^{-\lambda_1 t}. \tag{1.19}$$

Eq. 1.19 is a linear first-order differential equation of the form

$$\frac{dy}{dt} + Py = Q, \tag{1.20}$$

where P and Q are functions of t. We solve explicitly as follows. First we obtain an integrating factor e^I such that $\frac{dI}{dt} = P$ and $I = \int P dt$ using the chain rule:

$$\frac{d}{dt}(ye^I) = \frac{dy}{dt}e^I + ye^I\frac{dI}{dt} = \frac{dy}{dt}e^I + ye^I P = e^I(\frac{dy}{dt} + Py) \tag{1.21}$$

Clearly, Eq. 1.21 gives the product of the left hand-side of Eq. 1.20 with the integrating factor, e^I. Eq. 1.20 can be rendered thus:

$$\frac{d}{dt}(ye^I) = e^I(\frac{dy}{dt} + Py) = Qe^I \tag{1.22}$$

With Q and e^I now functions of t only, both sides of Eq. 1.22 can be integrated with respect to t giving:

$$ye^I = \int Qe^I dt + c. \tag{1.23a}$$

$$y = e^{-I} \int Qe^I dt + ce^{-I}. \tag{1.23b}$$

Applying Eqs.1.23a and 1.23b to Eq. 1.19, we see that P is λ_2 and

$$I = \int \lambda_2 dt = \lambda_2 t.$$

Also, $e^I = e^{\lambda_2 t}$ and $N_2 = y$. Of course, $Q = \lambda_1 N_0 e^{-\lambda_1 t}$. Therefore,

$$N_2 e^{\lambda_2 t} = \int \lambda_1 N_0 e^{-\lambda_1 t} e^{\lambda_2 t} dt + c = \lambda_1 N_0 \int e^{(\lambda_2 - \lambda_1)t} dt + c$$

$$= \frac{\lambda_1 N_0}{(\lambda_2 - \lambda_1)} e^{(\lambda_2 - \lambda_1)t} + c. \tag{1.24}$$

Applying boundary conditions to find c, we have that $N_2 = 0$ at $t = 0$ since we started out with pure radium at $t = 0$, thus,

$$0 = \frac{\lambda_1 N_0}{(\lambda_2 - \lambda_1)} e^{(\lambda_2 - \lambda_1)0} + c.$$

So

$$c = -\frac{\lambda_1 N_0}{(\lambda_2 - \lambda_1)}.$$

Finally, we obtain N_2 in Eq. 1.24 by substituting the above expression for c and dividing by $e^{\lambda_2 t}$:

$$\boxed{N_2(t) = \frac{\lambda_1 N_0}{(\lambda_2 - \lambda_1)} (e^{-\lambda_1 t} - e^{-\lambda_2 t}).} \tag{1.25}$$

Ernest Rutherford went further and solved the differential equation for any second product (polonium in our example). A quick look at Eq. 1.18 allows one to write similar equations for the rate of decay of N_3:

$$\frac{dN_3}{dt} - \lambda_2 N_2 - \lambda_3 N_3 \tag{1.26a}$$

$$\frac{dN_3}{dt} = \lambda_2 N_2. \tag{1.26b}$$

Eq. 1.26a is for N_3 unstable or still decaying while Eq. 1.26b is for N_3 stable. The same strategy done for N_2 can be applied to obtain a solution for N_3 as a stable third product, noting as well that $N_3 = 0$ at $t = 0$:

$$\boxed{N_3(t) = N_0 \lambda_1 \lambda_2 \left[\frac{e^{-\lambda_1 t}}{(\lambda_2 - \lambda_1)(\lambda_3 - \lambda_1)} + \frac{e^{-\lambda_2 t}}{(\lambda_1 - \lambda_2)(\lambda_3 - \lambda_2)} + \frac{e^{-\lambda_3 t}}{(\lambda_1 - \lambda_3)(\lambda_2 - \lambda_3)}. \right]}$$
$$\tag{1.27}$$

1.2.1 The Bateman Equations

In his paper of 1910, "Solution of a system of differential equations occurring in the theory of radioactive transformations" [6], Bateman observed that Rutherford did not go beyond solving for N_3 and that doing so based on the differential equations and steps outlined above was laborious and unsymmetrical. He found a general solution by taking Laplace transforms for the variables in the differential equations for N_1, N_2, N_3... along with complex integrals. There are very interesting alternative ways of solving this system of differential equations (see [24], [81]), such as the recent matrix exponential approach [68]. Using our notations in this book, here are the main steps taken by Bateman:

1.2.1.1 The System of Differential Equations

The amounts of radioactive material and their different products in a decay chain vary according to the system of differential equations:

$$\frac{dN_1}{dt} = -\lambda_1 N_1. \tag{1.28a}$$

$$\frac{dN_2}{dt} = \lambda_1 N_1 - \lambda_2 N_2. \tag{1.28b}$$

$$\frac{dN_3}{dt} = \lambda_2 N_2 - \lambda_3 N_3. \tag{1.28c}$$

$$\frac{dN_4}{dt} = \lambda_3 N_3 - \lambda_4 N_4. \tag{1.28d}$$

where N_1, N_2, N_3, N_4 give the number of atoms of the primary nuclide and successive products present at time t. An obvious generalization for any successive product at time t is:

$$\frac{dN_k}{dt} = \lambda_{k-1} N_{k-1} - \lambda_k N_k. \tag{1.29}$$

where λ_k is the decay constant of the kth nuclide.

1.2.1.2 Laplace Transformations

By definition, the Laplace transform [105] of $N(t)$ is given by

$$\mathscr{L}\{N(t)\} = \check{N}(s) = \int_0^\infty e^{-st} N(t)dt, \tag{1.30}$$

where s is a complex number parametrizing frequency

$$s = \sigma + i\omega.$$

Assuming initial conditions, $N_1(t=0) = N_1(0)$, $N_k(t=0) = 0$ for $k \geq 1$, and applying the Laplace transformation to both sides of Eq. 1.28a and Eq. 1.29, we obtain:

$$\int_0^\infty e^{-st}\frac{dN_1}{dt}dt = \int_0^\infty e^{-st}dN_1 = e^{-st}N_1(t)\Big|_0^\infty + s\int_0^\infty e^{-st}N_1(t)dt$$
$$= -N_1(0) + s\check{N}_1(s) = -\lambda_1\check{N}_1(s). \qquad (1.31a)$$

$$\int_0^\infty e^{-st}\frac{dN_k}{dt}dt = \int_0^\infty e^{-st}dN_k = e^{-st}N_k(t)\Big|_0^\infty + s\int_0^\infty e^{-st}N_k(t)dt$$
$$= s\check{N}_k(s) = \lambda_{k-1}\check{N}_{k-1}(s) - \lambda_k\check{N}_k(s) \qquad (1.31b)$$

Now algebraic, we solve for $\check{N}_1(s)$ and $\check{N}_k(s)$ at once in Eq. 1.31a and Eq. 1.31b:

$$\check{N}_1(s) = \frac{N_1(0)}{\lambda_1 + s} \qquad (1.32a)$$

$$\check{N}_k(s) = \frac{\lambda_{k-1}}{\lambda_k + s}\check{N}_{k-1}(s) = \frac{\lambda_1\lambda_2...\lambda_{k-1}}{(\lambda_1+s)(\lambda_2+s)...(\lambda_k+s)}N_{10}. \qquad (1.32b)$$

Exercise 10

1. Obtain $\check{N}_2(s)$ in preparation for getting $N_2(t)$.

2. Obtain $\check{N}_3(s)$ in preparation for getting $N_3(t)$.

1.2.1.3 Inverse Laplace Transformations or Partial Fractions

Bateman explicitly obtained the $N_k(t)$ using partial fractions and also showed a scheme for doing so using inverse Laplace transformations involving the so-called Bromwich integral [95]. Partial fractions for Eq. 1.32b gives:

$$\check{N}_k(s) = \frac{c_1}{(\lambda_1+s)} + \frac{c_2}{(\lambda_2+s)} + ...\frac{c_k}{(\lambda_k+s)}, \qquad (1.33)$$

where the c_k, called Bateman constants, are given by:

$$c_1 = \frac{\lambda_1 \lambda_2 ... \lambda_{k-1}}{(\lambda_2 - \lambda_1)(\lambda_3 - \lambda_1)...(\lambda_k - \lambda_1)} N_1(0), \qquad (1.34)$$

$$c_2 = \frac{\lambda_1 \lambda_2 ... \lambda_{k-1}}{(\lambda_1 - \lambda_2)(\lambda_3 - \lambda_2)...(\lambda_k - \lambda_2)} N_1(0), \qquad (1.35)$$

and so on, leading to a general expression for the Bateman constants, c_k:

$$c_k = \frac{\lambda_1 \lambda_2 ... \lambda_{k-1}}{(\lambda_1 - \lambda_k)(\lambda_3 - \lambda_k)...(\lambda_{k-1} - \lambda_k)} N_1(0). \qquad (1.36)$$

In view of Eq. 1.30, we know that

$$\frac{1}{\lambda + s} = \int_0^\infty e^{-st} e^{-\lambda t} dt, \qquad (1.37)$$

hence, $N_k(t)$ is obtained from $\check{N}_k(s)$ using

$$N_k(t) = c_1 e^{-\lambda_1 t} + c_2 e^{-\lambda_2 t} + \cdots c_k e^{-\lambda_k t}. \qquad (1.38)$$

Substituting for the c_i in Eq. 1.34 and Eq. 1.35, we have, finally,

$$N_k(t) = \frac{\lambda_1 \lambda_2 ... \lambda_{k-1}}{(\lambda_2 - \lambda_1)(\lambda_3 - \lambda_1 ...(\lambda_k - \lambda_1)} N_1(0) e^{-\lambda_1 t}$$
$$+ \frac{\lambda_1 \lambda_2 ... \lambda_{k-1}}{(\lambda_1 - \lambda_2)(\lambda_3 - \lambda_2 ...(\lambda_k - \lambda_2)} N_1(0) e^{-\lambda_2 t} + \cdots c_k e^{-\lambda_k t}. \qquad (1.39)$$

The need for summation over products of partial fractions is now obvious in Eq. 1.39. Hence, Eq. 1.39 is usually tied together as:

$$N_k(t) = \lambda_{k-1} \lambda_{k-2} ... \lambda_1 N_1(0) \sum_{j=1}^{k} \left(\frac{e^{-\lambda_j t}}{\prod_{i=1(i \neq j)}^{k} (\lambda_i - \lambda_j)} \right). \qquad (1.40)$$

To make Eq. 1.40 more intelligible and useful for calculations, we express the number $N_k(t)$ of nuclei remaining at generation k as a simple sum by tying Eq. 1.38, thus:

$$N_k(t) = c_1 e^{-\lambda_1 t} + c_2 e^{-\lambda_2 t} + ... c_k e^{-\lambda_k t} = \sum_{j=1}^{k} c_j e^{-\lambda_j t}. \qquad (1.41)$$

Likewise, we express the constants c_j or c_k, (that is, the c_1 of Eq. 1.34, c_2 of Eq. 1.35 and so on) in condensed product form thus:

$$c_k = N_1(0) \frac{\prod_{i=1}^{k-1} \lambda_i}{\prod_{i=1(i \neq j)}^{k} (\lambda_i - \lambda_j)}. \tag{1.42}$$

Finally, inserting Eq. 1.42 into 1.41, we obtain a compact and easy-to-compute expression for $N_k(t)$ thus:

$$\boxed{N_k(t) = N_1(0) \sum_{j=1}^{k} \left[\frac{\prod_{i=1}^{k-1} \lambda_i}{\prod_{i=1(i \neq j)}^{k} (\lambda_i - \lambda_j)} \right] e^{-\lambda_j t}.} \tag{1.43}$$

Example 1.4 *Use the Bateman equations to calculate the number of nuclei remaining in (a) the first generation with $k = 1$ and (b) the second generation, for a radioactive decay series at $t \geq 0$. For this series, let $t = 0$ be the defining initial conditions, that is, $N_1(t = 0) = N_1(0)$, $N_2(t = 0) = N_3(t = 0) = \cdots = N_k(t = 0) = 0$.*

Solution

(a) For the first generation with $k = 1$, $N_1(t)$ can be obtained using Eq. 1.42 and Eq. 1.41:

$$c_1 = N_1(0) \frac{\prod_{i=1}^{k-1} \lambda_i}{\prod_{i=1(i \neq j)}^{k} (\lambda_i - \lambda_j)} = (1) \times N_1(0) = N_1(0). \tag{1.44}$$

So

$$N_1(t) = \sum_{j=1}^{k=1} c_j e^{-\lambda_j t} = c_1 e^{-\lambda_1 t} = N_1(0) e^{-\lambda_1 t}, \tag{1.45}$$

as expected. (b) For the second generation with $k = 2$, the daughter nuclei produced, $N_2(t)$ can be obtained using Eq. 1.42 and Eq. 1.41:

$$c_1 = N_1(0) \frac{\prod_{i=1}^{2-1} \lambda_i}{\prod_{i=1(i \neq 1)}^{k} (\lambda_i - \lambda_1)} = N_1(0) \frac{\lambda_1}{\lambda_2 - \lambda_1}. \tag{1.46}$$

$$c_2 = N_1(0) \frac{\prod_{i=1}^{2-1} \lambda_i}{\prod_{i=1(i \neq 2)}^{k} (\lambda_i - \lambda_2)} = N_1(0) \frac{\lambda_1}{\lambda_1 - \lambda_2}. \tag{1.47}$$

Therefore,

$$N_2(t) = \sum_{j=1}^{k=2} c_j e^{-\lambda_j t} = c_1 e^{-\lambda_1 t} + c_2 e^{-\lambda_2 t} = N_1(0)\frac{\lambda_1}{\lambda_2 - \lambda_1}(e^{-\lambda_1 t} - e^{-\lambda_2 t}),$$

(1.48)

which is the expected result as in Eq. 1.25.

Example 1.5 *For a radioactive decay series with* $t = 0$ *being the defining initial conditions, that is,* $N_1(t = 0) = N_1(0)$, $N_2(t = 0) = N_3(t = 0) = ... = N_k(t = 0) = 0$, *confirm explicitly that the third generation of products with* $k = 3$ *for this series at* $t \geq 0$ *is given by (using Eq. 1.42 and Eq. 1.41):*

$$c_1 = N_1(0)\frac{\prod_{i=1}^{3-1}\lambda_i}{\prod_{i=1(i\neq 1)}^{3}(\lambda_i - \lambda_1)} = N_1(0)\frac{\lambda_1\lambda_2}{(\lambda_2 - \lambda_1)(\lambda_3 - \lambda_1)},$$

(1.49)

$$c_2 = N_1(0)\frac{\prod_{i=1}^{3-1}\lambda_i}{\prod_{i=1(i\neq 2)}^{3}(\lambda_i - \lambda_2)} = N_1(0)\frac{\lambda_1\lambda_2}{(\lambda_1 - \lambda_2)(\lambda_3 - \lambda_2)},$$

(1.50)

$$c_3 = N_1(0)\frac{\prod_{i=1}^{3-1}\lambda_i}{\prod_{i=1(i\neq 3)}^{3}(\lambda_i - \lambda_3)} = N_1(0)\frac{\lambda_1\lambda_2}{(\lambda_1 - \lambda_3)(\lambda_2 - \lambda_3)}.$$

(1.51)

Solution

Since,

$$N_3(t) = \sum_{j=1}^{k=3} c_j e^{-\lambda_j t} = c_1 e^{-\lambda_1 t} + c_2 e^{-\lambda_2 t} + c_3 e^{-\lambda_3 t},$$

(1.52)

the explicit value for $N_3(t)$ using the results of Eq. 1.49, Eq. 1.50 and Eq. 1.51 is given by:

$$N_3(t) = N_1(0)\lambda_1\lambda_2$$

$$\times \left[\frac{e^{-\lambda_1 t}}{(\lambda_2 - \lambda_1)(\lambda_3 - \lambda_1)} + \frac{e^{-\lambda_2 t}}{(\lambda_1 - \lambda_2)(\lambda_3 - \lambda_2)} + \frac{e^{-\lambda_3 t}}{(\lambda_1 - \lambda_3)(\lambda_2 - \lambda_3)}.\right]$$

(1.53)

Exercise 11

Consider a radioactive decay series with parent A as N_1 decaying down to the fourth generation N_4 in the manner:

$$A \rightarrow B \rightarrow C \rightarrow D.$$

Let $t = 0$ be the defining initial conditions, that is, $N_1(t = 0) = N_1(0)$, $N_2(t = 0) = N_3(t = 0) = N_4(t = 0) = 0$. Now carefully confirm that the fourth generation of products with $k = 4$ for this series at $t \geq 0$ is given as follows, using Eq. 1.42 and Eq. 1.41:

$$c_1 = N_1(0)\frac{\prod_{i=1}^{4-1}\lambda_i}{\prod_{i=1(i\neq 1)}^{4}(\lambda_i - \lambda_1)} = N_1(0)\frac{\lambda_1\lambda_2\lambda_3}{(\lambda_2 - \lambda_1)(\lambda_3 - \lambda_1)(\lambda_4 - \lambda_1)}, \quad (1.54)$$

$$c_2 = N_1(0)\frac{\prod_{i=1}^{4-1}\lambda_i}{\prod_{i=1(i\neq 2)}^{4}(\lambda_i - \lambda_2)} = N_1(0)\frac{\lambda_1\lambda_2\lambda_3}{(\lambda_1 - \lambda_2)(\lambda_3 - \lambda_2)(\lambda_4 - \lambda_2)}, \quad (1.55)$$

$$c_3 = N_1(0)\frac{\prod_{i=1}^{4-1}\lambda_i}{\prod_{i=1(i\neq 1)}^{4}(\lambda_i - \lambda_3)} = N_1(0)\frac{\lambda_1\lambda_2\lambda_3}{(\lambda_1 - \lambda_3)(\lambda_2 - \lambda_3)(\lambda_4 - \lambda_3)}, \quad (1.56)$$

$$c_4 = N_1(0)\frac{\prod_{i=1}^{4-1}\lambda_i}{\prod_{i=1(i\neq 4)}^{4}(\lambda_i - \lambda_4)} = N_1(0)\frac{\lambda_1\lambda_2\lambda_3}{(\lambda_1 - \lambda_4)(\lambda_2 - \lambda_4)(\lambda_3 - \lambda_4)}. \quad (1.57)$$

Therefore, find $N_4(t)$.

1.3 TRANSIENT AND SECULAR EQUILIBRIA

As one can deduce from Figure 1.1, radioactive equilibrium exists when daughter nuclide decays at the same rate at which it is being produced. This can happen between any pair in a decay series whenever $\lambda_{k-1}N_{k-1} = \lambda_k N_k$ provided the half life of parent nuclide $T_{\frac{1}{2}k-1}$ is greater than that of the daughter $T_{\frac{1}{2}k}$. In secular equilibrium, the parent nuclide has an extremely long half-life compared to the daughter nuclide. A good example of secular equilibrium occurs naturally with Uranium 238 having $T_{\frac{1}{2}} = 4.5$ billion years in secular equilibrium with Thorium 234 ($T_{\frac{1}{2}} - 24.1$ days). In transient equilibrium the half-life of the parent nuclide is not only longer than that of the daughter nucleus, but also the concentration of parent nuclide significantly decreases in time. The parent and daughter nuclides may decay at essentially the same rate, but the concentrations of both nuclides decrease as the concentration of the parent nuclide decreases. These features of transient and secular equilibria can be seen graphically in the Fig. 1.1. Note that in transient equilibrium, the activity of the daughter A_2 exceeds that of the parent A_1 by a factor that depends on the ratio of their half-lives. No equilibrium is obtained when $T_{1/2}$ of the daughter is greater than that of the parent. An important application of secular equilibrium in

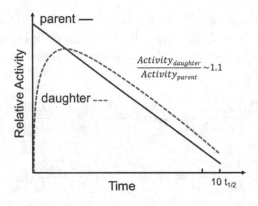

(a) Transient equilibrium.

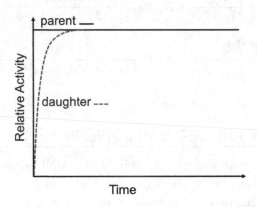

(b) Secular equilibrium.

Figure 1.1 Applications of Bateman Equations. The time is plotted in units of duaghter half-lives. Secular equilibrium is essentially achieved in about $7\,T_{1/2}$.

medical physics and nuclear medicine is the case of radium needles used in brachytherapy.

Example 1.6 *Show how the $T_{1/2}$ of a long-lived parent nuclide can be measured within a short time using relations arising from secular equilibrium.*

Solution

In secular equilibrium, $T_{1/2}$ of parent is much greater than that of daughter. Therefore,

$$\lambda_1 << \lambda_2.$$

The daughter nuclide rapidly builds up to a maximum (see Fig. 1.1), establishing equilibrium, with

$$N_1\lambda_1 = N_2\lambda_2.$$

This implies that in secular equilibrium, parent activity A_1 becomes equal or approximately equal to daughter activity A_2:

$$A_1 \approx A_2. \tag{1.60}$$

Hence,

$$\lambda_2 = \frac{N_1\lambda_1}{N_2}$$

and the required $T_{1/2}$ of the parent is obtained from this λ_2 using:

$$T_{1/2} = \frac{\ln 2}{\lambda_2} = \frac{(\ln 2)N_2}{N_1\lambda_1}.$$

Example 1.7 *Consider the expression for the daughter product in Eq. 1.25. Derive a specific relation for the relationship between parent and daughter activities for isotopes in transient equilibrium.*

Solution

In transient equilibrium, $T_{1/2}$ of parent is only slightly greater than that of daughter or even just about the same as that of the daughter nuclide. Hence,

$$\lambda_1 < \lambda_2.$$

The daughter activity starts from zero, rises to a maximum and continues decaying at approximately the same rate as that of the parent (see Fig. 1.1). To obtain a specific relation for the activity of the daughter, we multiply both sides of Eq. 1.25 with λ_2:

$$N_2(t)\lambda_2 = \frac{\lambda_2\lambda_1 N_0}{(\lambda_2 - \lambda_1)}(e^{-\lambda_1 t} - e^{-\lambda_2 t}) \tag{1.61}$$

Since $\lambda_1 < \lambda_2$, after sufficient time elapses, we get:

$$e^{-\lambda_2 t} << e^{-\lambda_1 t}$$

and Eq. 1.61 becomes

$$N_2(t)\lambda_2 = \frac{\lambda_2\lambda_1 N_0}{(\lambda_2 - \lambda_1)}e^{-\lambda_1 t}. \tag{1.62}$$

Using Eq. 1.17b, we get the required special relation for transient equilibrium:

$$N_2(t)\lambda_2 = \frac{\lambda_2\lambda_1 N_1}{(\lambda_2 - \lambda_1)}, \tag{1.63}$$

or in terms of activity A,

$$\boxed{A_2(t) = \frac{\lambda_2}{(\lambda_2 - \lambda_1)}A_1.} \tag{1.64}$$

1.3.1 Transient Equilibrium Applications

The compound decay process applied in a Technetium-99m[1] generator which is widely used for diagnostic procedures in nuclear medicine produces Technetium-99m (Tc-99m) from Molybdenum-99 (Mo-99). Tc-99m is a gamma-emitting radionuclide and this makes it useful for Single Photon Emission Computed Tomography (SPECT), for instance. The half-life of Tc-99m is just 6 hours and it would be difficult to store it for use when needed and it would be hard to transport it to hospital sites. Hence, for nuclear medicine purposes, Mo-99 is used to produce Technetium-99m when needed. The half-life of Mo-99 is 67 hours. Hence, the ratio of the half-lives of parent to daughter is $67/6 = 11.167$. The two isotopes illustrate application of transient equilibrium (see Fig. 1.2). The decay constant λ for Mo-99 is less than the decay constant for Tc-99m. However, the actual rate of decay of Mo-99 is initially larger than that of Tc-99m because of the great difference in their initial concentrations. As the concentration of the daughter or product Tc-99m, increases, the rate of decay of the daughter will approach and eventually be equal to the decay rate of the parent. When this occurs, the parent and daughter are said to be in transient equilibrium. With Mo-99 and Tc-99m, this happens after about four half-lives of the daughter. In the Tc-99m generator, four half-lives should be 24 hours and in practice, transient equilibrium is observed in 23 hours.

---

[1]The "m" refers to "meta-stable".

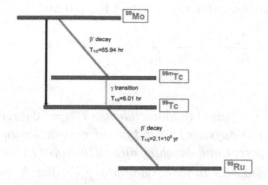

(a) Decay Scheme of Mo-99

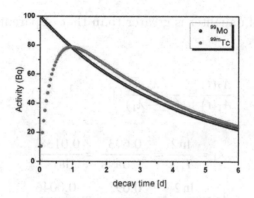

(b) Transient Equilibrium with Tc-99m

Figure 1.2 Decay Scheme of Mo-99 and Transient Equilibrium with Tc-99 m [22]

Exercise 12

Just as the ratio of daughter activity A_2 to parent activity A_1 in transient equilibrium has been shown to be (see Eq. 1.64)

$$\frac{A_2(t)}{A_1(t)} = \frac{\lambda_2}{(\lambda_2 - \lambda_1)}, \tag{1.65}$$

and that for secular equilibrium has been shown to be (see Eq. 1.60):

$$\frac{A_2(t)}{A_1(t)} \approx 1, \tag{1.66}$$

show that this ratio for the non-equilibrium case (which happens when

$T_{1/2}$ of daughter is greater than that of the parent) is:

$$\frac{A_2(t)}{A_1(t)} = \frac{\lambda_2}{(\lambda_2 - \lambda_1)}(1 - e^{-(\lambda_2 - \lambda_1)t}). \tag{1.67}$$

Example 1.8 *If a parent radionuclide has $T_{1/2} = 50$ days and the daughter has $T_{1/2} = 150$ days, (a) what type of radioactive equilibrium is expected between parent and daughter after 210 days (7 months)? (b) What is the ratio of daughter to parent activity $\frac{A_2(t)}{A_1(t)}$ after 210 days?*

Solution

(a). Here, $T_{1/2}$ of daughter is greater than that of parent, so it is a non-equilibrium case.
(b). Using Eq. 1.67,

$$\frac{A_2(t)}{A_1(t)} = \frac{\lambda_2}{(\lambda_2 - \lambda_1)}(1 - e^{-(\lambda_2 - \lambda_1)t}).$$

Here,

$$\lambda_1 = \frac{\ln 2}{T_{1/2}} = \frac{0.693}{50 \text{ days}} = \frac{0.0138}{\text{days}}.$$

$$\lambda_2 = \frac{\ln 2}{T_{1/2}} = \frac{0.693}{150 \text{ days}} = \frac{0.0046}{\text{days}}.$$

Hence,

$$\frac{A_2(t)}{A_1(t)} = \frac{0.0046}{(0.0046 - 0.0138)}(1 - e^{-(0.0046 - 0.0138) \times 210}) = 2.9517 \approx 3.$$

As expected, the activity of the daughter gets higher than that of the parent since the parent with a shorter half-life has decayed faster, producing more daughter nuclei for the latter's higher activity.

1.3.2   Matrix Exponential and Other Methods for Bateman Equations

Edmond Levy has recently demonstrated how the Bateman solution to the radioactive decay equations can be derived using the matrix exponential function [68]. Beginning with the system of differential equations for decay chains which we recall here for convenience,

$$\frac{dN_1}{dt} = -\lambda_1 N_1, \tag{1.28a revisited}$$

$$\frac{dN_2}{dt} = \lambda_1 N_1 - \lambda_2 N_2, \qquad\qquad \text{(1.28b revisited)}$$

$$\frac{dN_3}{dt} = \lambda_2 N_2 - \lambda_3 N_3, \qquad\qquad \text{(1.28c revisited)}$$

$$\frac{dN_k}{dt} = \lambda_{k-1} N_{k-1} - \lambda_k N_k, \qquad\qquad \text{(1.29 revisited)}$$

and denoting the derivative $\frac{dN_k}{dt}$ as $\dot{N}_k$, he expressed Eqs. 1.28a, 1.28b, 1.28c and 1.29 in matrix form thus:

$$\begin{bmatrix} \dot{N}_1 \\ \dot{N}_2 \\ \dot{N}_3 \\ \vdots \\ \dot{N}_k \end{bmatrix} = \begin{bmatrix} -\lambda_1 & 0 & 0 & \cdots & 0 & 0 \\ \lambda_1 & -\lambda_2 & 0 & \cdots & 0 & 0 \\ 0 & \lambda_2 & -\lambda_3 & & & 0 \\ \vdots & & & \ddots & \ddots & \vdots \\ 0 & 0 & 0 & 0 & \lambda_{k-1} & -\lambda_k \end{bmatrix} \begin{bmatrix} N_1 \\ N_2 \\ N_3 \\ \vdots \\ N_k \end{bmatrix}$$

Taking the inverse of the diagonal elements, the matrix equation becomes:

$$\dot{N}_k = A N_k, \qquad\qquad (1.69)$$

with the matrix A containing the decay constants. Several methods are available for solving Eq. 1.69 including the algebraic method involving inverse matrices, eigenvectors and eigenvalues [81]. Levy then provides a novel matrix exponential approach in which the matrix equation, that is, Eq. 1.69 has the immediate and unique solution:

$$\boxed{N_k(t) = e^{At} N_0.} \qquad\qquad (1.70)$$

Levy then shows [68] how to find the elements of the $n \times n$ matrix exponential, e^{At}, both algebraically and computationally. The short code for the computational solution is available for the curious reader.

1.4   RADIOACTIVE DECAY ENERGY CALCULATIONS

In general, the Q-value of a nuclear reaction is the total amount of energy absorbed or released during the reaction. In radioactive decay, the Q-value is obviously positive since radioactivity seeks to bring stability to an unstable nucleus via the release of energy through the emission of ionizing particles and radiation. Using the principle of conservation of energy as well as energy-mass equivalence, Q can be obtained thus:

$$Q = K_f - K_i = (m_i - m_f) \times c^2, \qquad\qquad (1.71)$$

where K_f and K_i are the final and initial kinetic energies of the species with masses m_f and m_i, respectively. The decay energy Q can also be expressed in terms of the binding energies B of the nuclear species:

$$Q = B_f - B_i. \tag{1.72}$$

Eq. 1.72 entails the conservation of nucleon number. Combining Eq. 1.71 and Eq. 1.72, we see that the sum of binding energy and rest mass of reactant nuclei is equal to that of the products:

$$B_i + m_i c^2 = B_f + m_f c^2. \tag{1.73}$$

In Eq. 1.71 and Eq. 1.73, the masses of specific nucleons are used if the species are nucleons as in the next example below. Likewise, the nuclear masses of the species can be used if the species are themselves nuclei. However, it is easier to measure atomic masses than nuclear masses. Hence, atomic masses can be used for the parent and daughter species consistently, since the total mass of the electrons in the daughter products is subtracted out of the mass of electrons and nucleons in the parent atom. Likewise, the practical way to calculate the binding energy of a nucleus is given thus:

$$B = (N m_n + Z m_{1_H} - m_{atom}) c^2. \tag{1.74}$$

where N is the number of neutrons in the nucleus, m_n the mass of the neutron, Z, the number of protons in the nucleus, m_{1_H} the mass of a neutral hydrogen atom (that is 1 proton and 1 electron) and m_{atom} is the mass of a neutral atom (including electrons) containing the nucleus of interest. For emphasis, the terms $Z m_{1_H}$ and m_{atom} each include the mass of Z electrons thereby cancelling out the electron mass so that we have the needed nuclear binding energy (of course, neglecting the electron binding energy which is usually about a million times smaller in comparison). For calculations, it is convenient to use MeV as unit for energy and MeV/c^2 for mass. In this case, c^2 and tabulated atomic masses are often given in terms of u or *amu*:

$$1\,u = 931.494 \frac{MeV}{c^2}$$

and

$$c^2 = 931.494 \frac{MeV}{u}.$$

For many practical purposes, the binding energy per nucleon $\frac{B}{A}$ is used and its formula is given by ([92]):

$$\frac{B}{A} = \frac{\Delta mc^2}{A} = \frac{Zm_pc^2 + (A-Z)m_nc^2 - Mc^2}{A}. \tag{1.75}$$

where Z and A are the atomic number and atomic mass number of the nuclide respectively, Mc^2 is the nuclear rest energy which can be obtained from nuclear data tables (see Appendix 4 or Atomic Weights and Isotopic Compositions from https://www.nist.gov/pml/atomic-weights-and-isotopic-compositions-relative-atomic-masses) either directly or by using the tables to calculate as follows, based on $A(u)$ as atomic mass in units of *amu* or *u* and electron mass m_e:

$$Mc^2 = A(u)c^2 - Zm_ec^2 = A(u) \times 931.494028\,\text{MeV/u} - Z \times 0.510999\,\text{MeV}, \tag{1.76}$$

again, neglecting the much smaller electron binding energy. In Eq. 1.75, m_pc^2 is the proton rest energy with value:

$$m_pc^2 = 938.272013\,\text{MeV};$$

m_nc^2, is the neutron rest energy, with value

$$m_nc^2 = 939.565346\,\text{MeV}.$$

Example 1.9 *What is the binding energy per nucleon $\frac{B}{A}$ for the Helium nucleus, He-4?*

Solution

From the Appendix or Atomic Weights and Isotopic Compositions, we get atomic mass of He-4 as **4.002603 u**. Hence, Eq. 1.76 gives the Helium nuclear rest energy:

$$Mc^2 = (4.002603\,u) \times (931.494028\,\text{MeV/u}) - 2 \times 0.510999\,\text{MeV}$$

which gives:

$$Mc^2 = 3728.400791\,\text{MeV} - 1.021998\,\text{MeV} = 3727.3788\,\text{MeV}.$$

Finally, Eq. 1.75 gives the binding energy per nucleon thus:

$$\frac{B}{A} = \frac{Zm_pc^2 + (A-Z)m_nc^2 - Mc^2}{A} \quad (1.77)$$

$$= \frac{2 \times (938.272013\,\text{MeV}) + 2 \times (939.565346\,\text{MeV}) - (3727.3788\,\text{MeV})}{4} \quad (1.78)$$

$$= 7.07398\,\text{MeV}. \quad (1.79)$$

The average binding energy per nucleon $\frac{B}{A}$ for a nucleus is a good measure of the cohesiveness of that nucleus. A graph of $\frac{B}{A}$ as a function of atomic mass number A is often plotted and used in explaining nuclear stability/instability. Such a graph enables empirical formulas for $\frac{B}{A}$ to be obtained from fitting.

Exercise 13

Models of the nucleus can be used to propose formulas for binding energy. One of such formulas, based on the liquid drop model of the nucleus is called the Weizsäcker empirical binding energy formula given thus:

$$\frac{B}{A}(^A_Z X) \approx C_1 A - C_2 A^{2/3} - C_3 \frac{Z^2}{A^{1/3}} - C_4 \frac{(A-2Z)^2}{A}, \quad (1.80)$$

where Z is the atomic number and A is the atomic mass number of nucleus X; the C_n are empirical constants such that C_1 accounts for nuclear volume effect, C_2 accounts for the nuclear surface effect, C_3 accounts for the electrostatic repulsion (Coulomb force) between protons, while C_4 accounts for the higher number of neutrons over protons in the nucleus. These constants have been empirically determined to be:

$$C_1 = 15.75\,\text{MeV},$$

$$C_2 = 17.80\,\text{MeV},$$

$$C_3 = 0.711\,\text{MeV},$$

$$C_4 = 23.70\,\text{MeV}.$$

Use the Weizsäcker empirical binding energy formula to calculate the binding energy of Cobalt 60 ($^{60}_{27}Co$) and Uranium 235 ($^{235}_{92}U$). Compare your result with the calculation using Eq. 1.74.

Example 1.10 *Outside the nucleus, free neutrons are unstable, with $T_{1/2} = 611 \pm 1\,s$ [10]. They decay into protons, electrons and electron antineutrinos, via the electroweak force, given by the following equations, including intermediary steps and at the quark level:*

$$n^0 \longrightarrow p^+ + W^- \longrightarrow p^+ + e^- + \bar{\nu}_e. \qquad (1.81)$$

$$udd \longrightarrow uud + W^- \longrightarrow uud + e^- + \bar{\nu}_e. \qquad (1.82)$$

Calculate the decay energy Q.

Solution

Both the nucleon level equation (Eq. 1.81) and the quark level equation (Eq. 1.82) describe the decay of the neutron via the emission of a W^- boson from one of the down quarks d of the neutron, thereby changing the down quark into an up quark u and the neutron into a proton in the intermediary step. In the final step, the W^- boson decays into the electron e^- and the electron antineutrino $\bar{\nu}_e$. Q is obtained using the reactants and the final products:

$$Q = (m_i - m_f) \times c^2 = (m_n - m_p - m_e - m_{\bar{\nu}_e}) \times c^2 = 0.782343\,\text{MeV}.$$

This Q is also the expected kinetic energy of the products since $K_i = K_n \approx 0$, thus, $Q = K_p + K_e + K_{\bar{\nu}_e}$. Routine measurements of Q gives values such as 0.782 ± 0.013 [49] wherein the precision is not sufficient to reveal the very small mass and kinetic energy of the neutrino. Note that whenever there are more than two products (as in this example), decay energy is divided among the products in a continuous distribution or continuous energy spectrum when measured.

1.5 MATHEMATICAL ELEMENTS OF ALPHA DECAY

1.5.1 Basics of Alpha Decay

There are more than $3{,}000$ known nuclides of which only 265 are stable, while the rest (over 90%) are radioactive [121]. This vast number of unstable nuclides seeks stability via the following processes: alpha decay, beta decay, gamma decay, electron capture and internal conversion. With respect to decay products besides the parent nuclides and daughter nuclides, the main products of nuclear decay are alpha particles, beta particles and gamma rays. It is standard practice to represent atomic nuclides (isotopes of elements) by referring to the chemical symbol, say X, and mass number (A) or to the atomic number (Z) and mass number (A). The number of neutrons is N. The mass number is the number of protons and neutrons. The atomic number is the number of protons. Hence $A = Z + N$. An alpha particle is a helium nucleus, with two protons and two neutrons, so $Z = 2$ and $A = 4$. Alpha decay is a quantum tunnelling process, mediated by the strong nuclear force and the electromagnetic force. The nuclear equation for alpha decay is given thus:

$$ {}_{Z}^{A}X \longrightarrow {}_{Z-2}^{A-4}Y + {}_{2}^{4}He(\alpha) \tag{1.83}$$

Note that the alpha particle escapes with most of the decay energy. For example, ^{238}U decays into ^{234}Th via alpha emission:

$$ {}_{92}^{238}U \rightarrow {}_{90}^{234}Th + {}_{2}^{4}He(K_\alpha = 4.2\,\text{MeV}) \tag{1.84}$$

Another example is the decay of ^{226}Ra into ^{222}Rn via alpha emission:

$$ {}_{88}^{226}Ra \rightarrow {}_{86}^{222}Rn + {}_{2}^{4}He(K_\alpha = 4.8\,\text{MeV}) \tag{1.85}$$

Example 1.11 *The nucleus ${}_{95}^{241}Am$ undergoes alpha decay:*

$$ {}_{95}^{241}Am \longrightarrow X + \alpha. \tag{1.86}$$

What is the atomic number Z and mass number A of the daughter nucleus, X?

Solution

For X, using Eq. 1.83, $Z = 95 - 2 = 93$ and $A = 241 - 4 = 237$.

Most of the decay energy Q becomes the kinetic energy of the alpha particle itself. Note that conservation of momentum entails that a small part of Q goes into the recoil of the nucleus. That is, the decay energy is divided into the kinetic energy of the alpha particle and the kinetic energy of the daughter nucleus, respectively:

$$ Q = K_\alpha + K_{Y_N}. \tag{1.87}$$

Hence, conservation of momentum entails that

$$ m_\alpha v_\alpha = m_Y v_Y, \tag{1.88}$$

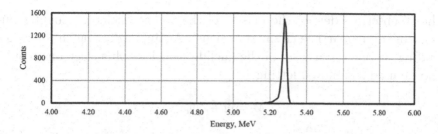

Figure 1.3 Plot of counts as a function energy showing the discrete energy spectrum for alpha decay. The single peak here is the alpha particle from Po-210 decay, with the 5.31 MeV peak measured as 5.29±0.05 MeV. Data was taken using a surface barrier detector by Creighton University NIM students, Fall 2019, and plotted by Zachary J. Sabata (Lab Partners: Jeffrey Wong, Alex Marvin and Caleb Thiegs).

where m_α and m_Y are the masses and v_α and v_Y are the velocities of the alpha particle and daughter nucleus, respectively. Consequently, as shown below in steps leading to Eq. 1.95, Eq. 1.87 becomes:

$$Q = K_\alpha \left(1 + \frac{m_\alpha}{m_Y}\right). \tag{1.89}$$

Since most radioisotopes that undergo alpha decay have mass number $A > 210$ and this is far greater than the $A = 4$ of the alpha particle, the part of the energy going to the recoil of the nucleus is generally quite small. In fact, K_α is usually only 2% smaller than Q. So for alpha decay,

$$Q \approx Q_\alpha$$

Hence, the alpha particle escapes with most of Q which can be detected as well-defined peaks with respect to energy spectrum. Emphatically, alpha decay has discrete energy spectrum, since the alpha particles escape with a definite energy or are monoenergetic, as shown in Figure 1.3. The alpha spectroscopy experiment leading to Figure 1.3 was performed using a surface barrier detector (ORTEC), CU-014-050-100 with an energy resolution of 14 keV FWHM). Of course, multiple discrete peaks are possible when there are multiple alpha decay modes as in a decay chain.

Alpha decay energy, Q_α, can be obtained from the general equation for decay energy Q, Eq. 1.71 as:

$$Q = (m_X - m_Y - m_\alpha) \times c^2, \tag{1.90}$$

where subscripts denote the masses of the parent nucleus, daughter nucleus and alpha particle (Helium nucleus). Again, neglecting the electron binding energy, the masses of the nuclides (nucleus plus electrons) could also be used as follows, letting,

$$M = m + Zm_e,$$

where Zm_e is the mass of Z electrons and m is the mass of Z protons and N neutrons (which can be obtained from $A = Z + N$. Thus,

$$Q_\alpha = (M_X - M_Y - M_\alpha) \times c^2, \tag{1.91}$$

Example 1.12 *Po-210 decays to a stable nuclide Pb-206 via alpha emission. Given the following decay equation, calculate the decay energy Q_α and the kinetic energy of the alpha particle T_α.*

$$^{210}_{84}Po \longrightarrow ^{206}_{82}Pb + \alpha. \tag{1.92}$$

Solution

(a) From Eq. 1.91,

$$Q_\alpha = (M_X - M_Y - M_\alpha) \times c^2$$

Applying this to Po-210 decay, we have,

$$Q_\alpha = (M_{Po} - M_{Pb} - M_\alpha - 2M_e) \times c^2,$$

where we are using the atomic masses (not just nuclear masses), in amu, of parents and daughters as well as accounting for the mass of the two orbital electrons lost during transition from Po 210 to Pb 206. Thus,

$$Q_\alpha = (210.0485 - 206.03883 - 4.00277 - (2 \times 0.00055)) \times c^2$$
$$= (0.0058\,\text{amu}) \times c^2$$

To obtain the result in MeV, we have

$$Q_\alpha = (0.0058\ \text{amu}) \times 931\,\text{MeV/amu} = 5.4\,\text{MeV}.$$

(b) The Q_α just calculated is shared as kinetic energy of the alpha particle and the recoiling daughter Pb-206 as follows:

$$Q_\alpha = T_\alpha + T_{Pb} = \frac{1}{2}M_{Pb}V_{Pb}^2 + \frac{1}{2}M_\alpha V_\alpha^2 \qquad (1.93)$$

Since, by conservation of linear momentum,

$$M_{Pb}V_{Pb} = M_\alpha V_\alpha,$$

$$V_{Pb} = \frac{M_\alpha V_\alpha}{M_{Pb}}.$$

Using $\frac{M_\alpha V_\alpha}{M_{Pb}}$ to substitute for V_{Pb} in Eq. 1.93, we have:

$$Q_\alpha = \frac{1}{2}M_{Pb}\frac{M_\alpha^2 V_\alpha^2}{M_{Pb}^2} + \frac{1}{2}M_\alpha V_\alpha^2 \qquad (1.94)$$

which gives,

$$Q_\alpha = T_\alpha(\frac{M_\alpha}{M_{Pb}} + 1) \qquad (1.95)$$

Thus,

$$T_\alpha = \frac{Q_\alpha}{1 + (\frac{M_\alpha}{M_{Pb}})} \qquad (1.96)$$

Substituting symbols with numbers,

$$T_\alpha = \frac{5.4\,\text{MeV}}{1 + (\frac{4}{206})} = 5.3\,\text{MeV}.$$

Example 1.13 *The primary branch of the alpha decay of Po-212 to the ground state of Pb-208 is part of the Thorium series. Given the following decay equation, calculate the decay energy Q_α and the kinetic energy of the alpha particle T_α. Use values from the Appendix or standard references including online: see National Institute of Standards and Technology (NIST) website, NIST.*

$$^{212}_{84}Po \longrightarrow ^{208}_{82}Pb + \alpha. \qquad (1.97)$$

Solution

$$Q_\alpha = (M_{Po} - M_{Pb} - M_\alpha - 2M_e) \times c^2 = 8.954 \text{ MeV}.$$

1.5.2 Geiger-Nuttall Law

Hans Geiger and John Mitchell Nuttall in 1911, formulated a rule now named after them, relating the decay constant of a radioisotope, λ, to the energy of the alpha particle it emits [40, 41].

$$log_{10}\lambda = -a_1\frac{Z}{\sqrt{E}} + a_2 \qquad (1.98)$$

where Z is the atomic number, a_1 and a_2 are supposedly constants (but actually coefficients characteristic of the isotopic series) and E is the kinetic energy of both the alpha particle and the daughter nucleus. Using the half-life and decay energy or Q-values of alpha decay Q_α, the Geiger-Nuttall law can be expressed [97] thus:

$$log_{10}T_{1/2} = a\sqrt{Q_\alpha} + b, \qquad (1.99)$$

where a and b are coefficients (constants) characteristic of the isotopic series. That is, the coefficients give rise to different lines for each isotope series. Since its initial formulation, numerous measurements on various isotopes have confirmed the validity of the Geiger-Nuttall law, and explanations of the underlying physics have been developed [96, 98]. One of the paradigm-shifting events that assured the acceptance of Quantum Mechanics was George Gamow's 1928 [39] use of quantum tunneling to explain Eq. 1.98, thus, giving a complete theory of alpha decay. Alternative formulations such as the use of time-independent matrix equations [117] confirmed Gamow's contribution. New mathematical generalizations of the Geiger-Nuttall law continue to be formulated in order to remove "constants" that are merely coefficients varying according to isotopes involved. Such generalizations have aimed at giving a universal decay law (UDL) [97]. In this vein, Qi, Liotta and Wyssa [97] generalize Eq. 1.98 thus:

$$log_{10}T_{1/2} = a\chi' + b\rho' + c, \qquad (1.100)$$

where χ' and ρ' are functions of the Q-values and atomic numbers of the nuclides involved.

Exercise 14

Which features below describe a typical alpha particle energy spectrum following detection?
A. One or more Gaussian-shaped peaks.   B. One or more Gaussian-shaped peaks, each with a lower energy continuous-Compton scatter background.   C. A continuous distribution.

1.6 MATHEMATICAL ASPECTS OF BETA DECAY

Beta decay is basically the emission of a positive electron (positron) or negative electron (an ordinary electron) along with the emission of a neutrino or anti-neutrino which occurs in nuclides in order to stabilize them with respect to the proportion of neutrons and protons.

1.6.1 Beta Decay Equations and Spectra

In Beta-minus (β^-) decay, a neutron is transformed into a proton, an electron β^- and an electron anti-neutrino, $\bar{v}_e$, via the weak nuclear force. The electron and the antineutrino share the available kinetic energy and are ejected from the nucleus. *Since the energy is shared between the electron and neutrino, the energy spectrum of beta decay is continuous, with an endpoint corresponding to Q* (see Fig. 1.4). The equation for this process is same as given in Eq. 1.81 and Eq. 1.82. The process can also be stated via Eq. 1.101:

$$_Z^A X \longrightarrow _{Z+1}^A Y + \beta^- + \bar{v}_e. \tag{1.101}$$

Eq. 1.101 can also be written as:

$$_Z^A X_N \longrightarrow _{Z+1}^A Y_{N-1} + \beta^- + \bar{v}_e. \tag{1.102}$$

For completeness, we state explicitly the *nuclear physics* and *particle physics* perspectives of the radiological equation (Eq. 1.101) for beta decay, respectively:

$$n \longrightarrow p + \beta^- + \bar{v}_e, \tag{1.103}$$

$$d \longrightarrow u + \beta^- + \bar{v}_e. \tag{1.104}$$

Using current understanding from elementary particle physics, the mechanism of β^- decay can be stated thus. A down quark d in a neutron, with electric charge $-\frac{1}{3}e$, frequently emits a negative charge $-e$, becoming an up quark u with charge $+\frac{2}{3}e$. Usually, u immediately reabsorbs the $-e$ and returns to its original state d. The W-minus boson, W^- bears the $-e$ that is quickly emitted and quickly reabsorbed. Since W^- is unstable, it can decay into an electron β^- and an electron anti-neutrino $\bar{v}_e$ (see Eq. 1.82).

An example of β^- decay is seen in the decay of ^{60}Co into an excited state of ^{60}Ni, with $T_{1/2} = 5.26$ *years*:

$$_{27}^{60} Co \longrightarrow _{28}^{60} Ni^* + \beta^- + \bar{v}_e. \tag{1.105}$$

Example 1.14 *One of the commonest products of the fission of Uranium-235, ^{235}U in nuclear reactors is Caesium-137, ^{137}Cs, also known as radiocaesium, with $T_{1/2} = 30.17\,years$. It is often used to calibrate radiation detection equipment because of its multiple decay modes: two β^- decays and one gamma decay in addition to other practical uses such as radiotherapy and thickness gauging. Make a standard sketch of the decay scheme of Cs-137 and explain briefly.*

Solution

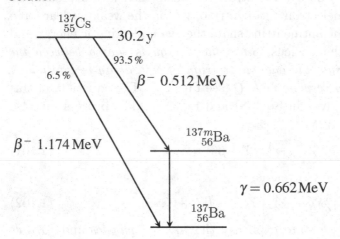

As can be seen from the decay scheme, about 93.5% decays by β^- emission to a metastable nuclear isomer of barium, denoted as barium-137m (Ba-137m). The rest (6.5%) directly decay into the ground state of barium-137, which is stable, still via another β^- emission. However, since the half-life of the metastable Ba-137m is merely about 153 seconds, it decays into the ground state via the emissions of gamma rays with energy 0.662 MeV. Gamma decay is discussed below. Interestingly, the gamma decay is observed only in 85% of the transformations. This implies that there is a mode of decay competing with the gamma decay. It is called internal conversion (discussed below). In the case of Cs-137, internal conversion competes with gamma decay for the 93.5% metastable Ba-137 nuclear state and therefore occurs in 8.5% of the transformations (93.5% − 85%).

A second mode of Beta decay is the *Beta-plus decay*, β^+. Here, a proton transforms into a neutron, leading to the production and release of a positron β^+ and an electron neutrino, ν_e. *The energy spectrum of beta decay is continuous* (see Fig. 1.4) because Q is shared between the positron and the neutrino. The equation for the process of β^+ decay is:

$$^A_Z X \longrightarrow ^A_{Z-1} Y + \beta^+ + v_e. \tag{1.106}$$

Eq. 1.106 can also be written as:

$$^A_Z X_N \longrightarrow ^A_{Z-1} Y_{N+1} + \beta^+ + v_e. \tag{1.107}$$

We can state explicitly the *nuclear physics* and *particle physics* perspectives of the radiological equation (Eq. 1.106) for beta-plus decay, respectively:

$$p \longrightarrow n + \beta^+ + v_e, \tag{1.108}$$

$$u \longrightarrow d + \beta^+ + v_e. \tag{1.109}$$

An example of β^+ decay is the decay of Nitrogen-13 into Carbon 13:

$$^{13}_7 N \longrightarrow ^{13}_6 C + \beta^+ + v_e. \tag{1.110}$$

Example 1.15 *The nucleus $^{65}_{30}Zn$ undergoes β^+ decay:*

$$^{65}_{30} Zn \longrightarrow X + \beta^+ + v_e. \tag{1.111}$$

What is the atomic number Z and mass number A of the daughter nucleus, X?

Solution

For X, using Eq. 1.106, $Z = 30 - 1 = 29$ and $A = 65$.

Exercise 15

Which features below describe a typical beta particle energy spectrum following detection?
A. One or more Gaussian-shaped peaks.   B. One or more Gaussian-shaped peaks, each with a lower energy continuous-Compton scatter background.   C. A continuous distribution.

Exercise 16

In β^+ decay, the parent and daughter nuclides are?
A. Isotopes (same number of protons Z).   B. Isotones (same number of neutrons $A - Z$).   C. Isobars (same mass number A).
D. Isomers (same mass number A and atomic number Z) but different energy states, eg, ground state versus excited state.

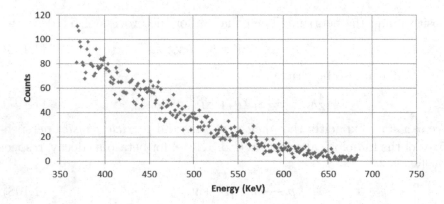

Figure 1.4 Counts versus energy showing the continuous energy spectrum for beta decay of Tl-204 to Pb-204 through beta-minus decay. The endpoint energy is not obvious from the plot owing to the asymptotic behavior towards the high energy end of the spectrum. Further analysis (Fermi-Kurie analysis) is usually done to determine the endpoint energy $Q = 764$ keV. Data was taken using a surface barrier detector by Creighton University NIM students, Fall 2017, and plotted by Aaron Herridge. Lab partners: Laura Aumen and Amrit Gautam.

1.6.2 Continuous Beta Spectrum and Neutrinos

The continuous spectrum from beta decay could not be explained in terms of only a recoiling daughter nucleus and an electron. If the electron were carrying away all the Q, there should be a discrete spectrum, as in alpha decay. Rather, as shown in Fig. 1.4, a continuous spectrum of possible and smaller energies for the electron is obtained. Back in 1930, no one could account for the missing Q. Niels Bohr went as far as ditching the principle of energy conservation in the beta decay process. To save the principles of energy conservation and momentum conservation, Wolfgang Pauli postulated in a letter to his colleagues (Lise Meitner and Hans Geiger), in 1930, the existence of an electrically-neutral, low mass particle that would be emitted along with the beta particle but hard to detect. He called it a "neutron". Enrico Fermi took up the postulation, renamed the hypothetical particle neutrino (Italian for *little neutral one*), after the neutron was discovered in 1932 by James Chadwick. Fermi worked out a theory of beta decay in which an electron-neutrino pair is spontaneously produced by a nucleus in a manner analogous to the way photons are spontaneously emitted by excited atoms. Fermi's theory led to the current understanding in terms of the theory of weak interaction, based on the weak nuclear force proposed in 1955 by Murray Gell Mann. Experimental evidence for neutrinos were presented by Reines and Cowan in 1956. The historical digression above was to motivate respect for mathematical physics and the reliability of conservation principles of physics. Particle physics, astrophysics, and astro-particle physics have grown enormously thanks to the postulation and discovery of neutrinos. We next outline Fermi's mathematico-physical theory of beta decay and the so-called Fermi-Kurie plot for extracting the endpoint energy Q from routine measurements of beta decay such as the one in Figure 1.4.

1.6.3 Transition Rate and Fermi-Kurie Plots

Enrico Fermi accounted for the elusive neutral particle, the neutrino, and sacrosanct principles of physics: the conservation of energy and of momentum in his theory of beta decay. Fermi proposed four fermions directly interacting with one another. Following rejection of his manuscript by the influential journal *Nature*, on the grounds of being largely speculative, Fermi submitted to Italian and German journals where his work was published in 1933/34 [35, 37]. He made extensive use of Dirac's work [29, 31] in quantum mechanics, specifically the latter's theory of radiation, an area he himself held some expertise [36]. A complete English translation was later made in 1968 [126]. Let us summarize the derivation of the so-called Fermi's 2nd golden rule which is used in the theory of beta decay. Note that Dirac derived the golden rule [31, 43] and Fermi used it extensively, making it famous by designating it another "golden" rule [36, 38]. Fermi's 2nd golden rule gives the transition rate $\Gamma_{a \to b}$ from an initial state a to a final state b. For a system that obeys the time-dependent Schrödinger equation,

$$i\hbar \frac{\partial |\Psi(r,t)\rangle}{\partial t} = H |\Psi(t)\rangle, \tag{1.112}$$

time-dependent perturbation theory (see Appendix A.3 for details) holds that for different initial and final states, denoted by subscripts a and b respectively,

$$\langle \psi_a | \psi \rangle = \delta_{ba} - \frac{i}{\hbar} \int_0^t \langle \phi_b | H'(t') | \phi_b \rangle \, e^{i\omega_{ba} t'} \, dt', \tag{1.113}$$

where the Hamiltonian $H'(t) = H_0 + V(t)$ and $V(t)$ is the perturbation. For weak perturbations and long times, $H'(t) = H'$ (constant in time), the transition probability $P_b(t)$ is given by

$$\Gamma_{a \to b} \times t, \; for \; b \neq a. \tag{1.114}$$

To obtain an expression for $\Gamma_{a \to b}$ we define $H'_{ba} \equiv \langle \phi_b | H' | \phi_a \rangle$ and $E_{ba} \equiv \hbar \omega_{ba}$. Several steps of algebra and integration over all possible states yield the probability that the system will be in any state different from the initial state:

$$P_{ba} = \lim_{t \to \infty} 4 |H'_{ba}|^2 \int_0^\infty \frac{\pi t}{2\hbar} \delta(E_{ba}) \rho(E_b) \, dE_b \tag{1.115}$$

$$= \lim_{t \to \infty} \frac{2\pi t}{\hbar} |H'_{ba}|^2 \int_0^\infty \rho(E_b) \delta(E_b - E_a) \, dE_b \tag{1.116}$$

Thus, the transition rate, obtained as probability per time, for going from a to any other state b becomes a constant given by:

$$\boxed{\Gamma_{a \to b} = \frac{dP_{ba}}{dt} = \frac{2\pi}{\hbar} |H'_{ba}|^2 \int_0^\infty \rho(E_b) \delta(E_b - E_a) \, dE_b} \tag{1.117}$$

Eq. 1.117 is called *Fermi's second golden rule*. Thanks to the Dirac delta function, the equation ensures that transitions are only possible if the initial and final states have the same energy. This is precisely the conservation of energy and momentum ingrained into the transition rate formalism. For more straightforward calculations, the transition rate can be presented only in terms of the perturbation V. This follows from the constraints we imposed on H', namely, that the perturbation is weak and that we wait long enough for $H'(t) = H'$ to be constant in time, hence,

$$V(t) = V$$

and Fermi's 2nd golden rule becomes:

$$\boxed{\Gamma_{a \to b} = \frac{2\pi}{\hbar} |\langle b|V|a \rangle|^2 \int_0^\infty \rho(E_b) \delta(E_b - E_a) dE_b,} \tag{1.118}$$

where the matrix element $|\langle b|V|a \rangle|$ is approximately constant within the energy range $|(E_b - E_a)| \leq 2\pi\hbar/t$ of the allowed final states b. Quantitatively, how weak should the perturbation V be? It is sufficiently weak when the total probability to reach any possible state satisfies

$$\sum_{b \neq a} P_b(t) = \sum_{b \neq a} \Gamma_{a \to b} \times t.$$

Now let us apply Fermi's 2nd golden rule to beta decay beginning with the following definitions. The initial state of the system is that of a neutron at rest, $|a\rangle$ with energy

$$E_a = m_n c^2 \tag{1.119}$$

The final state of the system is now a proton (at rest, relatively), a free electron with wave vector k_e and a free electron anti-neutrino with wave vector k_v (where $k = 2\pi/\lambda = p/\hbar$),

$$|b\rangle = |k_e, k_v\rangle$$

with energy,

$$E_b = E_p + E_e + E_v = m_p c^2 + \sqrt{(m_e c^2)^2 + (p_e c)^2} + \sqrt{(m_v c^2)^2 + (p_v c)^2}.$$

Thus,

$$E_b = m_p c^2 + \sqrt{m_e^2 c^4 + \hbar^2 c^2 k_e^2} + \sqrt{m_v^2 c^4 + \hbar^2 c^2 k_v^2}. \tag{1.120}$$

The transition matrix for beta decay, meets the conditions of Eq. 1.117 and so we state explicitly:

$$|\langle b|V|a\rangle|^2 = |\langle k_e, k_v|V_{weak}|a\rangle|^2 = constant \equiv g^2. \tag{1.121}$$

Fermi treated the transition shown in Eq. 1.121 as a point interaction [35]. As shown in Figure 1.4, measurements provide the energy and therefore momentum of the electron $p_e = \hbar k_e$. Hence, we evaluate the transition to a final state k_e with arbitrary anti neutrino momentum k_v:

$$\Gamma_{a \to k_e} = \frac{2\pi}{\hbar} g^2$$
$$\times \int_0^\infty \delta\left((m_p - m_n)c^2 + \sqrt{m_e^2 c^4 + p_e^2 c^2} + \sqrt{m_v^2 c^4 + \hbar^2 c^2 k_v^2}\right) d^3 k_v. \tag{1.122}$$

For convenience, let

$$(m_p - m_n)c^2 + \sqrt{m_e^2 c^4 + p_e^2 c^2} \equiv -A(p_e)$$

and

$$d^3 k_v \equiv 4\pi k_v^2 dk_v.$$

We employ the generalized scaling property of the Dirac delta function (for which g generalizes to $g(x)$):

$$\int_{-\infty}^{\infty} g(x)\delta(f(x))dx = \sum_{i} \frac{g(x_i)}{|f'(x_i)|} \tag{1.123}$$

in which the x_i satisfy $f(x_i) = 0$ for $-\infty < x < \infty$ as well as a step function:

$$\zeta(x) = \begin{cases} 1 & x > 0 \\ 0 & x < 0 \end{cases} \tag{1.124}$$

which give the result following algebraic simplifications:

$$\boxed{\Gamma_{a \to k_e} = \frac{8\pi}{\hbar^4 c^3} g^2 A(p_e)\sqrt{A(p_e)^2 - m_\nu^2 c^4}\, \zeta(A(p_e) - m_\nu c^2).} \tag{1.125}$$

Eq. 1.125 shows a peculiar momentum dependence close to the maximum electron momentum as expected (that is for small A). When Eq. 1.125 is fitted to high precision experimental results, an upper limit for neutrino mass can be obtained which is of the order of $\frac{2\,eV}{c^2}$ in which case, the kinetic energy of the daughter nucleus is no longer negligible.

It is hard to detect the neutrinos and it is clear why detected electron energy of beta decay gives a continuous energy spectrum as seen in Fig. 1.4. From such spectra, Q_β should be the endpoint of the counts versus energy graph. However, in practice, the count rate near the endpoint energy is small due to electronic noise and limited resolution of the detector. A better approximation of the Q_β is obtained by using Fermi theory outlined above to do a so-called Fermi-Kurie plot [46,65,70]. Q_β is proportional to the transition probability of Eq. 1.125. The Fermi-Kurie plot is a linear function which when extrapolated to intersect with the energy axis, gives the endpoint energy, Q_β. The Fermi-Kurie plot makes use of Fermi's theory of beta decay, to give a functional form for the plot of counts of electrons $N(T_e)$ as a function of energy. First Q_β is stated as the sum of the kinetic energy of the beta particle and the neutrino:

$$Q_\beta = T_e + T_\nu, \tag{1.126}$$

leading to:

$$E_e^2 = m_e^2 c^4 + p_e^2 c^2 = (m_e c^2 + T_e)^2. \tag{1.127}$$

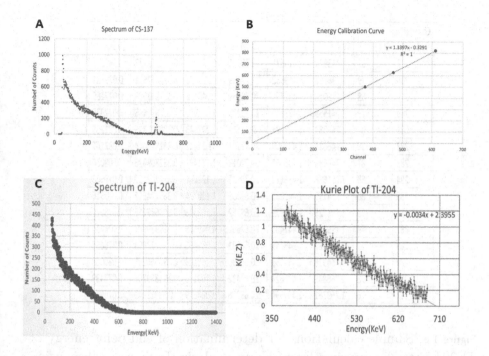

Figure 1.5 Stages in the determination of end-point energy of beta decay using Fermi-Kurie analysis and plots. A. Beta spectrum of Cs-137; the peaks between 600 keV and 700 keV are due to internal conversion (IC) electrons and these peaks are used for calibrating the system. B. Energy calibration curve plotted using IC peaks from (A) as well as from pulse generator. C. Beta spectrum of Tl-204. D. Fermi-Kurie plot for Tl-204, showing a linear fit whose x-axis intercept gives the end-point energy. Data was taken using a surface barrier detector by Creighton University NIM students, Fall 2017, and plotted by Amrit Gautam. Lab partners: Aaron Herridge and Laura Aumen.

$(Q_\beta - T_e)^2$ is proportional to the counts of electrons as a function of energy, $N(T_e)$. Thus, the functional form can be obtained:

$$N(T_e) = C\sqrt{(T_e)^2 + 2T_e m_e c^2}(Q_\beta - T_e)^2(T_e + m_e c^2)F(Z', T_e), \quad (1.128)$$

where $F(Z', T_e)$ is called the Fermi function for beta decay (which is different from the Fermi function for Fermi-Dirac statistics), introduced to correct for the Coulomb interaction between the beta particle from the

Tl-204

Channel Number	N(E)	W	P	G(Z,W)	K(Z,E)	Energy (MeV)	Error in K
277	81	1.733976	1.416571	20.78551	1.138464	0.375061737	0.069233
278	79	1.736502	1.419662	20.77459	1.12298	0.376352551	0.079381
279	68	1.739029	1.422751	20.76369	1.040628	0.377643589	0.077185
280	73	1.741555	1.425839	20.75279	1.076926	0.37893485	0.077975
281	92	1.744083	1.428924	20.7419	1.207543	0.380226335	0.081258
282	98	1.746611	1.432009	20.73101	1.24482	0.381518044	0.082147
283	87	1.749139	1.435091	20.72013	1.171491	0.382809976	0.08006
284	80	1.751668	1.438172	20.70925	1.122047	0.384102133	0.078663
285	80	1.754197	1.441252	20.69838	1.120724	0.385394512	0.078515
286	78	1.756726	1.444329	20.68752	1.105323	0.386687116	0.078012
287	70	1.759256	1.447405	20.67666	1.045875	0.387979943	0.076434
288	72	1.761787	1.45048	20.66581	1.059465	0.389272994	0.076656
289	92	1.764318	1.453553	20.65496	1.196203	0.390566269	0.080008
290	78	1.766849	1.456624	20.64412	1.100145	0.391859767	0.077439
291	76	1.769381	1.459694	20.63328	1.084679	0.39315349	0.076948

Figure 1.6 Sample calculations for determination of end-point energy of Tl-204 beta decay using Fermi-Kurie analysis. Data was taken using a surface barrier detector by Creighton University NIM students, Fall 2017, and analysed by Amrit Gautam. Lab partners: Aaron Herridge and Laura Aumen.

decay and the daughter nucleus. This nuclear Coulomb interaction shifts the distribution toward lower energies for beta-minus decay because of the attraction between the daughter nucleus and the emitted electron. It shifts the distribution towards higher energies for beta-plus emission. The Q_β can finally be obtained by plotting $Q_\beta - T_e$ against T_e. Eq. 1.128 can also be stated in terms of the momentum instead of the kinetic energy:

$$N(p_e) = C p_e^2 \left[Q_\beta - \sqrt{(p_e)^2 c^2 + m_e^2 c^4} + m_e c^2) \right] F(Z', p_e). \quad (1.129)$$

The Fermi function can be obtained from tables and is approximately given by:

$$F(Z', T_e) \approx \frac{2\pi\eta}{1 - e^{-2\pi\eta}} \quad (1.130)$$

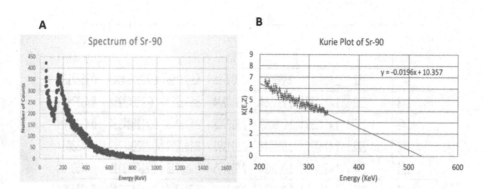

Figure 1.7 Determination of end-point energy of beta decay using Fermi-Kurie analysis and plots. A. Beta spectrum of Sr-90. B. Fermi-Kurie plot for Sr-90, showing a linear fit whose x-axis intercept gives the end-point energy of 528 ± 17 keV, which agrees with literature value of 546 keV. Data was taken using a surface barrier detector by Creighton University NIM students, Fall 2017, and plotted by Amrit Gautam. Lab partners: Aaron Herridge and Laura Aumen.

where Z' is the atomic number of the nucleus and $\eta = \pm \frac{Z'\alpha}{\beta}$. Note that here, $\beta = \frac{v}{c}$ and $\alpha = \frac{1}{137}$ is the fine structure constant. Fig. 1.5 shows aspects of the procedures in Fermi-Kurie analysis to get the end point energy. Essentially, the steps involved using Eq. 1.128, with the Fermi function, $F(Z', T_e)$, calculated for each selected channel in the form:

$$K(Z', T_e) = \frac{\sqrt{\frac{N(T_e)}{G(Z,W)}}}{W} \qquad (1.131)$$

where W is the total relativistic energy of the beta particle: $W = \frac{T_e}{m_0 c^2} + 1$, T_e is the kinetic energy of the beta particle and $G(Z,W)$ is the modified Fermi function often tabulated in terms of the relativistic momentum of the beta particle, p_e. For calculations without interpolations from data tables, one may calculate $K(Z', T_e)$ (Eq. 1.131 using):

$$G(Z,W) = \frac{F(Z', T_e)p}{W} \qquad (1.132)$$

Fig. 1.6 shows in tabular form the calculations done to obtain Fig. 1.5D. Note that $K(Z', T_e)$ in Eq. 1.131 is the same as $K(Z, E)$ in Fig. 1.5 and

Sr-90

Channel Number	N(E)	W	P	G(Z,W)	K(Z,E)	Energy (MeV)	Error in K
147	252	1.409317	0.993063	2.909738	6.603356	0.209160848	0.359218
148	252	1.411786	0.996564	2.910224	6.591257	0.21042258	0.459791
149	255	1.414255	1.000059	2.910708	6.618246	0.211684534	0.459566
150	222	1.416725	1.003549	2.911192	6.163897	0.212946713	0.43303
151	244	1.419196	1.007034	2.911675	6.450318	0.214209115	0.446996
152	235	1.421667	1.010513	2.912157	6.318714	0.215471741	0.438246
153	245	1.424138	1.013987	2.912639	6.440026	0.216734591	0.443208
154	226	1.42661	1.017456	2.913119	6.174046	0.217997665	0.42739
155	263	1.429082	1.020919	2.913599	6.648226	0.219260962	0.451227
156	213	1.431555	1.024377	2.914079	5.972157	0.220524483	0.413783
157	214	1.434028	1.027831	2.914557	5.975345	0.221788227	0.412522
158	221	1.436501	1.031279	2.915035	6.061334	0.223052196	0.415578
159	214	1.438975	1.034722	2.915512	5.953825	0.224316388	0.4086
160	198	1.44145	1.038161	2.915989	5.716631	0.225580804	0.395097
161	206	1.443925	1.041594	2.916465	5.820506	0.226845443	0.399056

Figure 1.8 Sample calculations for determination of end-point energy of Sr-90 beta decay using Fermi-Kurie analysis. Data was taken using a surface barrier detector by Creighton University NIM students, Fall 2017, and analysed by Amrit Gautam. Lab partners: Aaron Herridge and Laura Aumen.

Fig. 1.6. An important caveat is that even in the most painstaking Fermi-Kurie analysis, agreement between theory and experimental result seems to be good only near the upper limit of the beta energy spectrum, as admitted by Kurie [65]. As shown in Fig. 1.7, for the case of Sr-90, the Kurie analysis (sample data shown in Fig. 1.8) does yield acceptable results, as in this case where an endpoint energy of 528 ± 17 keV was obtained. Literature value for Sr-90 endpoint energy is 546 keV.

1.7   MATHEMATICAL PHYSICS OF GAMMA DECAY

1.7.1   Isomeric Transition Energetics and Multipole Selection Rules

Gamma decay, the emission of electromagnetic radiation of high frequency, is characterized by a change in energy without a change in atomic number Z or atomic mass A. The nuclear equation for gamma decay is given by:

$$\;^A_Z X^* \longrightarrow\; ^A_Z X + \gamma \qquad (1.133)$$

where $*$ indicates an excited state of the nucleus. Note that gamma transitions do not have to get to the ground state of the nucleus. The energies of the emitted γ rays can vary from a few keV to many MeV.

Gamma decay is described as an isomeric transition. Isomers here refers to two different nuclear configurations for the same isotope. Precisely, the isomers typically have different total angular momenta and always have different energies. Alpha and beta decays often leave nuclei in excited states. De-excitation of such states via gamma decay often follows or accompanies alpha and beta decays. Other processes of de-excitation besides gamma decay include internal-conversion electrons, and internal conversion from interaction between nucleus and extra-nuclear electrons. The gamma photon escapes from the nucleus and is detected via the photoelectric effect, Compton-effect and or pair production, depending on the available decay energy. This is why the energy spectrum of gamma decay usually consists of one or more Gaussian-shaped peaks each with a lower energy continuous-Compton scatter background, as can be seen in the case of Co-60 in Fig. 1.9. These and several other features (back scatter, x-ray fluorescence, escape peak, sum peak, annihilation peak, etc.) enable development of the theory of photon interactions with matter, mathematically explained further in Chapter 5.

Example 1.16 *The nucleus $^{60}_{28}Ni$ undergoes gamma decay:*

$$^{60}_{28}Ni \longrightarrow X + \gamma. \qquad (1.134)$$

What is the atomic number Z and mass number A of the daughter nucleus, X?

Solution

For X, using Eq. 1.133, $Z = 28$ and $A = 60$.

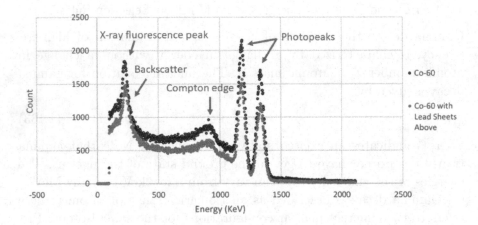

Figure 1.9 Counts versus energy showing the Gaussian-shaped peaks (photopeaks) from Co-60 decay with the expected 1173 keV and 1333 keV peaks. These peaks are followed by a lower energy continuous Compton scatter beginning with so-called Compton edge around 960 keV. Part of the experiment involved taking data with (red font) and without (black font) extra lead sheets between source and detector. Data was taken using a NaI(Tl) scintillation detector by Creighton University NIM students, Fall 2019, and plotted by Caleb Thiegs (Lab Partners: Jeffrey Wong, Alex Marvin and Zachary Sabata).

Since gamma decay arises from electromagnetic effects, it can be tracked with respect to changes in the charge and current distributions in nuclei. Electric moments are involved with charge distributions. Magnetic moments result from current distributions. Hence, gamma decays can be classified as electric (E) or magnetic (M) multipole radiations. The E and M multipole radiations differ in parity properties. As in beta decay, transition probabilities decrease rapidly as changes in angular-momentum (with quantum number L) increase. This leads to selection rules with various degrees of likelihood. These characteristics of gamma decay can be described mathematically as classical radiation of electromagnetic energy from a charged system. The radiation can be expanded in vector spherical harmonics, corresponding to the various multipoles of the charge and current distributions. The angular distribution of the radiation with respect to the radiating system takes a geometric shape that is uniquely determined by the order of the multipole (see Table 1.1). It turns out that the order of the multipole depends upon the angular momentum quantum numbers (L) of the photons, the spins (intrinsic angular momenta) of the initial (I_i) and final (I_f) nuclear states, as well as their parities, involved in the transition. Conservation of these system properties leads to the following vector equation, Eq. 1.135:

$$\vec{I}_i = \vec{I}_f + \vec{L}. \tag{1.135}$$

Eq. 1.135 engenders the spin selection rule that the magnitude of the photon angular momentum L is allowed any integer value between the sum and the difference of the intrinsic angular momentum quantum numbers of the initial and final states:

$$|I_i - I_f| \leq L \leq |I_i + I_f| \tag{1.136}$$

Next is the parity selection rule arising from the fact that parity is conserved to a very high degree in electromagnetic transitions such as gamma decay. Radiation fields have odd or even parity for a given L depending on their type. For electric multipole radiations, *E1*, *E2*, up to *EL* the parity is $(-1)^L$. For magnetic multipole radiations, *M1*, *M2*, up to *ML*, the parity is $(-1)^{L+1}$. E and M multipole radiations for a given L have opposite parity. An intuitive classical analogue to this quantum phenomenon is the difference between electric dipole moment, $\vec{p}_e = q\vec{r}$; and magnetic dipole moment which is proportional to $q\vec{r} \times \vec{v}$. Under parity inversion, E has odd parity since r transforms to $-r$ while M has even parity since $\vec{r} \times \vec{v}$ transforms to $-\vec{r} \times -\vec{v}$.

The following examples illustrate theory and applications of selection rules based on the aforementioned electromagnetic characteristics focusing on parity, spin and angular momentum.

1.7.1.1 Parity, Spin and Angular Momentum in Multipole Selection Rules

Example 1.17 *The parity operation or spatial inversion generated by the parity operator $\widehat{\Pi}$ in 3-dimensions is given by*

$$\widehat{\Pi}\psi(\vec{r}) = \psi'(\vec{r}) = \psi(-\vec{r}). \tag{1.137}$$

Show that the angular momentum operator $\hat{L}$ is even under parity transformation.

Solution

Since $\vec{L} = \vec{r} \times \vec{p}$, where p is momentum, then

$$\hat{L}' = \hat{\Pi}^\dagger \hat{L} \hat{\Pi} = (-\hat{r}) \times (-\hat{p}) = (\hat{r}) \times (\hat{p}) = \hat{L}. \qquad (1.138)$$

Example 1.18 *Generally as mentioned earlier, the parity of an electric multipole is given by*

$$\hat{\Pi}(EL) = (-1)^L. \qquad (1.139)$$

Calculate the parities of (a) electric dipole radiation (E1) and (b) electric quadrupole radiation (E2).

Solution

For (a)

$$\hat{\Pi}(E1) = (-1)^1 = -1. \qquad (1.140)$$

For (b)

$$\hat{\Pi}(E2) = (-1)^2 = +1. \qquad (1.141)$$

Example 1.19 *In general, the parity of a magnetic multipole is given by*

$$\hat{\Pi}(ML) = (-1)^{L+1} = -(-1)^L. \qquad (1.142)$$

Calculate the parities of (a) magnetic dipole radiation (M1) and (b) magnetic quadrupole radiation (M2).

Solution

For (a)

$$\hat{\Pi}(M1) = (-1)^{1+1} = +1. \qquad (1.143)$$

For (b)

$$\hat{\Pi}(M2) = (-1)^{2+1} = (-1)^3 = -1. \qquad (1.144)$$

Example 1.20 *Table 1.1 shows characteristics of multipolarity as well as angular distribution using Legendre polynomials. Use Table 1.1 and the selection rule (Eq. 1.136) to determine the multipolarity of the following transitions from initial to final nuclear spins and parities (+ or*

Table 1.1 Charactaristics of Multipoles

L	Multipolarity	$\Pi(EL)$	$\Pi(ML)$	angular distribution, $x \equiv \cos\theta$
1	dipole	-1	+1	$P_2(x) = \frac{1}{2}(3x^2 - 1)$
2	quadrupole	+1	-1	$P_4(x) = \frac{1}{8}(35x^4 - 30x^2 + 3)$
3	octupole	-1	+1	$P_6(x) = \frac{1}{16}(231x^6 - 315x^4 + 105x^2 - 5)$
4	hexadecapole	+1	-1	$P_8(x) = \frac{1}{128}(6435x^8 - 12012x^6$
				$+6930x^4 - 1260x^2 + 35)$
$\vdots$ $\vdots$		$\vdots$	$\vdots$	$\vdots$

-):

(a) $2^+ \rightarrow 1^-$

(b) $2^+ \rightarrow 1^+$

(c) $3^+ \rightarrow 1^-$

(d) $3^+ \rightarrow 1^+$

Solution

(a) $2^+ \rightarrow 1^-$. From Eq. 1.136, L can only be 1. There is a change in parity $(+ \rightarrow -)$. In the $L = 1$ row of Table 1.1, a parity change (denoted by -1) is in the electric transition (EL) column. This is an E1 transition.

(b) $2^+ \rightarrow 1^+$. From Eq. 1.136, L can only be 1. There is no change in parity $(+ \rightarrow +)$. In the $L = 1$ row of Table 1.1, no parity change (denoted by +1) is in the magnetic transition (ML) column. This is an M1 transition.

(c) $3^+ \rightarrow 1^-$. There are two possible transitions. From Eq. 1.136, L can be 1 or 2. There is a change in parity $(+ \rightarrow -)$. In the $L = 1$ row of Table 1.1, a parity change (denoted by -1) is in the electric transition (EL) column. This corresponds to an E1 transition. In the $L = 2$ row of Table 1.1, a parity change (denoted by -1) is in the magnetic transition (ML) column. This corresponds to an M2 transition. The transition with the largest change in L will dominate. This will be primarily an M2 transition.

(d) $3^+ \rightarrow 1^+$. There are two possible transitions. From Eq. 1.136, L can be 1 or 2. There is no change in parity $(+ \rightarrow +)$. In the $L = 1$ row of Table 1.1, no parity change (denoted by +1) is in the electric transition (ML) column. This corresponds to an M1 transition. In the $L = 2$ row of Table 1.1, no parity change (denoted by +1) is in the electric transition (EL) column. This corresponds to an E2 transition. The transition with the largest change in L will dominate. This will be primarily an E2 transition.

Example 1.21 *Table 1.1 shows characteristics of multipolarity as well as angular distribution using Legendre polynomials. Use Table 1.1 and the selection rule (Eq. 1.136) to state the possible values of L as well as possible types of emitted radiation for the following transition from initial to final nuclear spin and parity:* $3^- \to 2^+$.

Solution

For the transition $3^- \to 2^+$, L ranges from 1 to 5. Possible radiations, considering change in parity are E1, M2, E3, M4, E5. The electric dipole transition (E1) will dominate.

Exercise 17

Which features below describe a typical gamma decay energy spectrum following detection?
A. One or more Gaussian-shaped peaks.   B. One or more Gaussian-shaped peaks, each with a lower energy continuous-Compton scatter background.   C. A continuous distribution.

1.7.2 Internal Conversion Coefficients

Competing with gamma decay is the process of *internal conversion* (IC) whereby the nuclear excitation energy is transferred to a K-shell orbital electron which is then ejected with a kinetic energy that is the difference between the excitation energy and the orbital electron binding energy. The ejected electrons can be detected. Figure 1.10 shows internal conversion electrons for Cs-137 and such spectra are important for calibration purposes in NIM. The beta particle spectroscopy experiment which required calibration with IC electrons of Figure 1.10 was performed using a surface barrier detector (ORTEC). The resulting vacancy in the K-shell is filled by higher level orbital electron (from L or M shells, for instance) and the transition energy is emitted as characteristic photons (Characteristic X-rays) or Auger electrons. The equation for internal conversion is:

$$_{Z}^{A}X^* \longrightarrow _{Z}^{A}X + e_k^-$$ (1.145)

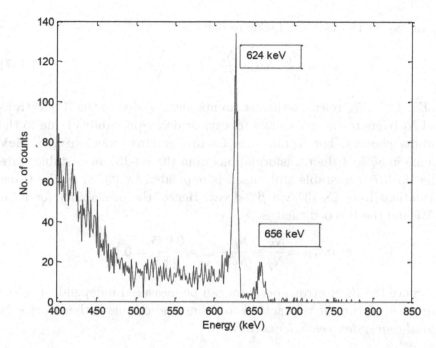

Figure 1.10 Counts versus energy showing the discrete energy spectrum for internal conversion electrons from Cs-137 decay with the expected 624 keV and 656 keV peaks. Data was taken using a surface barrier detector by Creighton University NIM students, Fall 2019, and plotted by Shani Perera (Lab Partner: Michael Mimlitz).

The energy of the internal conversion electron E_{IC} that is detected is

$$E_{IC} = E_{nuc^*} - E_b, \qquad (1.140)$$

where E_{nuc^*} is the nuclear excitation energy and E_b is the binding energy in the electron shell. Note that it is the electron binding energy of the daughter nuclide that is used in Eq. 1.146, not that of the parent, whenever the transformation is from a metastable daughter nuclide instead of an unstable parent. The probability for internal conversion is greatest for the K-shell electrons and decreases rapidly with the outer shells. This leads to the use of internal conversion coefficients to denote the probability of K-, L- and even M-shell conversions. Since the internal conversion competes with the gamma decay, the internal conversion coefficient α_K,

α_L or α_M for the various shells are defined thus:

$$\alpha_K = \frac{N_K}{N_\gamma}, \quad \alpha_L = \frac{N_L}{N_\gamma}, \quad \alpha_M = \frac{N_M}{N_\gamma}... \tag{1.147}$$

In Eq. 1.147, N_K refers to the net counts (decays) due to the IC electrons and N_γ refers to the net counts (decays or decay probability) due to the gamma photons. For instance, in Cs-137, gamma emission $0.662\,\mathrm{MeV}$ occurs in 85% of the transformations from the Ba-137 metastable state. The Ba-137 metastable state itself is populated by 93.5% of the transformations from Cs-137 via β^- decay. Hence, the probability for IC is 8.5% and the IC coefficient is

$$\alpha_{total} = \frac{N_K}{N_\gamma} + \frac{N_L}{N_\gamma} + ... = \frac{0.085}{0.85} = 0.1.$$

Since the IC electron spectrum can be measured independent of the gamma spectrum, it is practical to define and calculate the *relative internal conversion coefficients* thus:

$$\frac{\alpha_K}{\alpha_L} = \frac{N_K}{N_L},... \tag{1.148}$$

where α_K and N_K are respectively the probability of internal conversion of K-shell electron and net number of counts due to the K-shell electron detected.

Example 1.22 *Having measured the IC spectrum of Cs-137, students in the NIM course at Creighton University obtained the following counts for the IC peaks: $N_K = 852 \pm 29$ for the 624 keV K-shell electron and $N_L = 147 \pm 12$ for the 656 keV L-shell electron (see Figure 1.10). What is the relative internal conversion coefficient for these IC electrons of Cs-137?*

Solution

Using Eq. 1.148,
$$\frac{\alpha_K}{\alpha_L} = \frac{N_K}{N_L} = \frac{852}{147} = 5.8 \pm 0.5$$

This result is in agreement with the literature (about 5.43) [122] and data tables:http://nucleardata.nuclear.lu.se/toi/nuclide..

Exercise 18

The K-shell binding energy of Ba-137 is 37 keV. What is the energy of the K-shell internal conversion electron? Hint: Consider the 662 keV gamma decay over which internal conversion competes about 5% of the time.

Exercise 19

A radioactive sample is a mixture of an unknown element A with another radioactive element B that has the following electron binding energies. K = 35 keV; L = 5 keV; M = 1.5 keV. Using a scintillation detector, you record two characteristic X-rays from the sample with energies 33.5 keV and 3.7 keV. Which of these X-rays, if any, is from element B?

1.7.3 Electron Capture versus Isomeric Transition

As the name implies, Electron Capture is a process whereby the nucleus captures one of its own K-shell electrons leading to the transformation of a proton into a neutron and the release of an electron neutrino. It is also called K-capture since the captured electron usually comes from the K-shell of the atom. Sometimes, there is L-capture from the L-shell. The equation for electron capture is as follows:

$$\,^A_Z X + e^-_k \longrightarrow \,^A_{Z-1} Y + \nu_e \qquad (1.149)$$

Eq. 1.149 can also be written as:

$$\,^A_Z X_N \longrightarrow \,^A_{Z-1} Y_{N+1} + \nu_e. \qquad (1.150)$$

More explicitly the *nuclear physics* and *particle physics* perspectives of the radiological equation (Eq. 1.149) for electron capture are, respectively:

$$p + e^- \longrightarrow n + \nu_e, \qquad (1.151)$$

$$u + e^- \longrightarrow d + \nu_e. \qquad (1.152)$$

Note that the single emitted neutrino carries the entire decay energy, giving electron capture a characteristic energy or single peak. Furthermore, the K-shell vacancy created is filled with higher level orbital electrons just as in internal conversion. The transition energy gives rise to characteristic X-ray photons. First observed by Luis Alvarez in Vanadium-48, important examples include:

$$\begin{array}{c}^{125}_{53}I + e^-_k \longrightarrow ^{125}_{52}Te^* + \nu_e\end{array}$$

$$\begin{array}{c}^{26}_{13}Al + e^-_k \longrightarrow ^{26}_{12}Mg^* + \nu_e\end{array}$$

$$\begin{array}{c}^{59}_{28}Ni + e^-_k \longrightarrow ^{59}_{27}Co^* + \nu_e\end{array}$$

$$\begin{array}{c}^{57}_{27}Co + e^-_k \longrightarrow ^{57}_{26}Fe^* + \nu_e\end{array}$$

1.7.4 Auger Electrons

Analogous to internal conversion where nuclear excitation energy is transferred to an orbital electron instead of being emitted as gamma rays, an excited atom can emit another atomic electron instead of emitting characteristic X-rays. The emitted electron is named an Auger electron and the process is termed *Auger effect*. Essentially, electrons with characteristic energies and typically belonging to outer shells, are ejected from atoms in response to a downward transition by another electron in the atom, typically an inner shell electron. Note that both electron capture and internal conversion can result in an inner shell vacancy as could a particle colliding with an inner electron. Any process that leaves a vacancy in an inner electron shell can give rise to the emission of Auger electrons. As a competing process with photon emission, measurements of Auger electrons could yield information about X-ray emissions and vice versa. The X-ray emission is often quantified as fluorescence yield, that is, the number of X-rays emitted per electron vacancy created.

Example 1.23 *Since the fluorescence yield ω of a given shell (K, L or M) is the number of fluorescence (characteristic) photons emitted per vacancy in the shell, ω can also be regarded as the probability, after creation of an electronic shell vacancy, of fluorescence photon emission as opposed to Auger electron emission What is the probability for Auger effect in terms of ω_K, ω_L and ω_M?*

Solution

The probability for Auger effect after creation of a vacancy in an electronic shell except the outer shell of the absorber atom is:

$$1 - \omega_K,$$

$$1 - \omega_L,$$

$$1 - \omega_M,$$

for K, L and M shells respectively.

1.7.5 Coincidences and Angular Correlations

Many nuclear processes produce two gamma rays or photons simultaneously, while other processes produce two or more photons in quick succession. The temporal and angular correlations between the two photons are often studied by setting up a coincidence detector system. Such coincidence detector systems have far reaching applications including identification of nuclear processes, decay modes, decay properties, absolute activity measurements, computerized tomography (CT) and positron emission tomography (PET). Consider two gamma rays, γ_1 and γ_2. Angular correlation denotes the probability of γ_2 being emitted at an angle relative to γ_1. When two or more gamma rays are emitted in cascade, a coincidence unit can be used to gain knowledge of the angular momentum distribution and multipolarity of the nuclear states involved in the transition. For instance, using two NaI(Tl) scintillation counters where the height of the electronic output pulses is proportional to the incident gamma ray energies, pulse height selection is done in a so-called single channel analyzer mode and then one counter is used to record γ_1 and the other γ_2.

Example 1.24 *Coincidence units can be used to determine the decay constant of radionuclides with extremely short half-lives (say nanoseconds). The apparatus is also the basis for Positron Annihilation Spectroscopy (PAS) which in turn has medical applications in Positron Emission Tomography (PET). Briefly describe the nuclear processes involved in PET and the role of coincidence techniques therein, including a diagram to illustrate such description.*

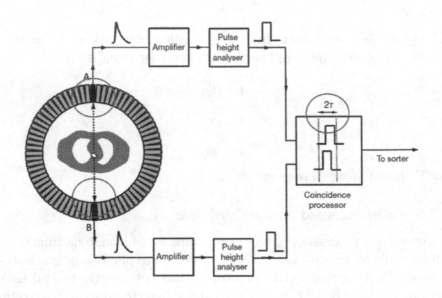

Figure 1.11 Coincidence Unit for PET scanning of a patient in transverse or cross-sectional view. Oppositely aligned detectors detect two photons associated with each other (from positron–electron annihilation) based upon their arrival times. Photons detected at detectors A and B produce signals that are amplified and analysed to determine whether they meet the energy acceptance criteria. Signals within the energy acceptance window produce a logic pulse of width τ which is sent to the coincidence processor. When both logic pulses fall within a specified interval (here, 2τ), a coincidence event is indicated.

Adapted from [52], p.370.

Solution

- PET is a nuclear medicine imaging technique that can reveal physiological states and functions based on metabolism and is useful in early cancer detection. During PET imaging, a positron-emitting radionuclide tagged with a biological marker is introduced into the body by injection or inhalation. The radionuclide circulates through the bloodstream to particular organs. The positrons emitted by the radionuclide have a very short range in tissue ($\approx 1\,\mathrm{mm}$) and undergo annihilation with an available atomic orbital electron typically at the end of their tracks, leading to the emission two gamma photons each with energy of 0.511 MeV, moving away from the point of production in nearly opposite directions for the conservation of angular momentum.

- Coincidence units including oppositely aligned detectors as shown in Fig. 1.11 pick up the angularly correlated photons based on a timing window of the order of few nanoseconds. Typical detectors are scintillators arranged in a ring about the patient.

- To pick up angularly correlated events, signals within the energy acceptance window produce a logic pulse of width τ which is sent to the coincidence processor. When both logic pulses fall within a specified interval (here, 2τ), a coincidence event is triggered and recorded

- The line connecting pairs of aligned detectors triggered in coincidence is called the coincidence line and many such lines are formed during acquisition of a PET scan. These lines give the spatial activity distribution in the organ of interest within the field-of-view.

- Mathematical algorithms are then applied to the spatially distributed coincidence lines to reconstruct the required images of internal organs of interest.

1.8 SPONTANEOUS FISSION

To account for the structure and therefore characteristics of the atomic nucleus, theoretical models have been developed. Three examples of nuclear models [64] are the *liquid drop model* (the nucleons behave like the molecules in a drop of liquid), *the shell model* (emphasizes the orbits of individual nucleons in the nucleus) and *the collective model* (accounts for whole-nuclear motion in the shell model. The shell model is useful in describing the excited states of nuclei. The liquid drop model predicts the possibility of spontaneous fission within a time short enough for observation by currently available methods when

$$\frac{Z^2}{A} \geq 47. \tag{1.153}$$

In Eq. 1.153, Z is the atomic number and A is the atomic mass. Since nuclei that undergo spontaneous fission also are subject to alpha decay, perhaps the quantum mechanical explanation of alpha decay might be extended to explain and model nuclear fission. Many transuranium elements exhibit spontaneous fission with the release of neutrons and the fission fragments. The latter are often unstable and so decay by β and γ radiation. The energy spectrum of the neutrons released is continuous up to a certain limit (about 10 MeV) and exhibits a Maxwellian shape (Maxwellian distribution is described in chapter 2).

Exercise 20

The nucleus $^{137}_{55}Cs$ undergoes beta decay:

$$^{137}_{55}Cs \longrightarrow X + \beta^- + \bar{v}_e. \tag{1.154}$$

What is the atomic number Z and mass number A of the daughter nucleus X?

Exercise 21

Radioactive Pd-103 or I-125 seeds are used to treat prostate cancers. For Pd-103 with a half-life of 17 days, what fraction (in %) of the original activity remains after 7 days of delivery of a consignment to a radiation therapy facility?

Exercise 22

The most frequently used positron emitters for PET scanning in nuclear medicine are: carbon-11 (C-11), nitrogen-13 (N-13), oxygen-15 (O-15), fluorine-18 (F-18) and rubidium-82 (Rb-82). Calculate the Q-value or β^+ decay energy Q_β for C-11, N-13, O-15, F-18 and Rb-82. Hint: Use any of Eq. 1.71, Eq. 1.73 for the Beta-plus decay process (Eq. 1.106), that is, use atomic rest energy method or nuclear rest energy method.

1.9 ANSWERS

Answer of exercise 1

(a) Expected is $0.07\,\mu$Ci and measured is $0.05\,\mu$Ci. (b) The sample may not be very pure. The detector is missing some counts perhaps coming from opposite side of sample.

Answer of exercise 2

(a) $\lambda = 0.132$ per year (b) $N(t)/No = 0.138$ or 13.8 % (c) For 98% to decay, it means 2% remains. So $N(t)/No = 0.02$ and $t = 29.6$ years.

Answer of exercise 3

70 mCi

Answer of exercise 4

Solution: $T_{1/2} = \frac{\ln 2}{(\lambda_1 + ...)}$ In about 89.28% of events, it decays to calcium-40 (40Ca) with emission of a beta particle (−, an electron) with a maximum energy of 1.31 MeV and an antineutrino. In about 10.72% of events, it decays to argon-40 (40Ar) by electron capture (EC), with the emission of a neutrino and then a 1.460 MeV gamma ray. The radioactive decay of this particular isotope explains the large abundance of argon (nearly 1%) in the Earth's atmosphere, as well as prevalence of 40Ar over other isotopes. Very rarely (0.001% of events), it will decay to 40Ar by emitting a positron (+) and a neutrino.

Answer of exercise 5

Ans $C\,e^{-\lambda t}$

Answer of exercise 6

$A(after\ 45\ mins) = A_0 e^{((-0.693/109.7)\times 45min)} = 0.75A_0 = 0.75 \times 740\,Bq = 555\,MBq$

Answer of exercise 7

$3.52 \times 10^{12}\ atoms$

Answer of exercise 8

Given in the text.

Answer of exercise 9

1 hour.

Answer of exercise 10

Straightforward algebra.

Answer of exercise 11

Straightforward algebra.

$$N_4(t) = \sum_{j=1}^{k=4} c_j e^{-\lambda_j t} = c_1 e^{-\lambda_1 t} + c_2 e^{-\lambda_2 t} + c_3 e^{-\lambda_3 t} + c_4 e^{-\lambda_4 t}, \qquad (1.58)$$

so explicit value for $N_4(t)$ using the results of Eq. 1.54, Eq. 1.55, Eq. 1.56 and Eq. 1.57 is given by:

$$N_4(t) = N_1(0)\lambda_1\lambda_2\lambda_3 \left[\frac{e^{-\lambda_1 t}}{(\lambda_2 - \lambda_1)(\lambda_3 - \lambda_1)(\lambda_4 - \lambda_1)} \right.$$
$$+ \frac{e^{-\lambda_2 t}}{(\lambda_1 - \lambda_2)(\lambda_3 - \lambda_2)(\lambda_4 - \lambda_2)}$$
$$+ \frac{e^{-\lambda_3 t}}{(\lambda_1 - \lambda_3)(\lambda_2 - \lambda_3)(\lambda_4 - \lambda_3)}$$
$$\left. + \frac{e^{-\lambda_4 t}}{(\lambda_1 - \lambda_4)(\lambda_2 - \lambda_4)(\lambda_3 - \lambda_4)} \right]. \quad (1.59)$$

Answer of exercise 12

Straightforward algebra.

Answer of exercise 13

For Co-60, 525 MeV, for U-235, 1782.9 MeV. Using analytic equation is close to empirical.

Answer of exercise 14

Ans A One or more Gaussian-shaped peaks.

Answer of exercise 15

Ans C. A continuous distribution.

Answer of exercise 16

Ans C. Mass number remains the same.

Answer of exercise 17

Ans B. One or more Gaussian-shaped peaks, each with a lower energy continuous-Compton scatter background.

Answer of exercise 18

Ans $0.662 \text{ MeV} - 0.037 \text{ MeV} = 0.625 \text{ MeV}$ which is close to that detected in the figure as 624 keV.

Answer of exercise 19

The 33.5 keV X-ray is from element B, precisely the M shell.

Answer of exercise 20

$Z = 56$, $A = 137$

Answer of exercise 21

Ans 75% of the original A_0.

Answer of exercise 22

Ans C-11 = 0.96 MeV. N-13 = 1.198 MeV. O-15 = 1.732 MeV. F-18 = 0.634 MeV.

CHAPTER 2

Probability and Statistics for Nuclear Experimental Data

Probability provides the theoretical expectation for experimental data while statistics encompasses the collection, organization, analysis, display and interpretation of data. Both are growing areas of research. In this chapter, we (I and you the reader) give definitions and mathematical characteristics of probability functions in view of the meaningful analysis of nuclear experimental data. We alert ourselves to limitations in the definitions and characterization and such alerts can be useful not only in ensuring scientifically valid interpretation of nuclear experimental data but also useful in spurring new extensions and discoveries in the probability/statistical theories themselves. For instance, by rigorously analysing confidence intervals in Poisson distributions with background noise (common in nuclear experiments using detectors) and Gaussian errors with bounded physical regions, Feldman and Cousins [34] have provided a new and widely utilized unified method for construction of classical confidence intervals [54], the Feldman-Cousins Confidence Intervals.

2.1 PROBABILITY DISTRIBUTIONS AND THEIR CHARACTERISTICS

A probability distribution is a mathematical construct that specifies the relative likelihood of all possible outcomes. Let each outcome of a random process such as radioactive disintegration be represented by a random variable x which ranges over all admissible values in the process. $P(x)$ is the probability density function which gives the expected frequency of occurrence for each possible outcome. For instance, if a die is thrown once, getting a "2" is one of six possible outcomes. Thus, for $x = 2$, $P(x) = 1/6$. Depending on the process, a random variable may be either **discrete** or **continuous**. For casting a die, x is discrete. So for discrete variables only a finite or denumerably infinite number (corresponding to natural numbers) of values is allowed. A random variable is continuous, if it takes a continuous range of values. If x is discrete, $P(x_i)$ then gives the frequency at each point x_i. If x is continuous, however, this interpretation is not possible because when continuous the variable can take on infinitely many, uncountable values between any interval and only probabilities of finding x in finite intervals have meaning. The distribution $P(x)$ is thus a continuous density function such that the probability of finding x between the interval x and $x + dx$ is $P(x)dx$. Think of the distinction between probability density and probability as analogous to the distinction between mass density and mass.

2.1.1 Cumulative Distribution

Since $P(x)dx$ is the probability of obtaining the outcome within dx, we denote as cumulative probability distribution the likelihood of finding x between any arbitrary interval, say x_1 and x_2, thus,

$$P(x_1 \leq x \leq x_2) = \int_{x_1}^{x_2} P(x)dx, \qquad (2.1)$$

for $P(x)$ continuous. For $P(x)$ discrete, the cumulative distribution is a sum:

$$P(x_1 \leq x \leq x_2) = \sum_{i=1}^{2} P(x_i). \qquad (2.2)$$

Since probability has to do with all given possibilities, total probability is given as one (unity) and mathematically, this means that the probability distribution is normalized to one:

$$\int_{-\infty}^{\infty} P(x)dx = 1 \qquad (2.3)$$

for continuous distribution and

$$P(x_1 \leq x \leq x_2) = \sum_{i=1}^{2} P(x_i) = 1, \qquad (2.4)$$

for discrete distribution. Our aim is to go from probability to statistics. $P(x)$ as a probability function has predictive value. It allows us to predict the distributions of experimental outcomes based on parameters of theory or hypothesis under test. This is especially important where and when it is impractical to obtain multiple measurements and one must be satisfied with only one measurement, for instance, during radionuclide imaging and nuclear measurements on patients [52]. We perform statistics on actual measured data and compare with the predictions of probability. Often, the comparison engenders the test of theory. Hence, we define some statistical measures next.

2.1.2 Expectation Values, Mean, Variance and Covariance

Consistent with probability being predictive, the expectation value $E(x)$ of a random variable x with a known distribution $P(x)$ is given by:

$$E(x) = \int_{-\infty}^{\infty} xP(x)dx, \tag{2.5}$$

when x is continuous. For the discrete case, we have:

$$E(x) = X = \sum_{i=1}^{n} x_i P(x_i). \tag{2.6}$$

Why is the expectation value the expected value? Well, the expected value is expected because after so many repetitions of the same experiment, it is the average value you get in the long-run. It can be predicted.

Example 2.1 *Find the expectation value $E(x)$ for the rolling of a six-sided die.*

Solution

We solve this using Eq. 2.6:

$$E(x) = (1 \times \frac{1}{6}) + (2 \times \frac{1}{6}) + (3 \times \frac{1}{6}) + (4 \times \frac{1}{6}) + (5 \times \frac{1}{6}) + (6 \times \frac{1}{6}) = 3.5$$

In actual experiments of rolling the die, the average of all the outcomes increasingly approaches 3.5 as the number of rolls approaches infinity. It will be closer to 3.5 if one rolls 500 times than if one roles 50 times. This leads us to the arithmetic mean and variance. The so-called law of large numbers has it that the arithmetic mean of the outcomes of experiments approximately converges to the expected value as the number of repetitions approaches infinity. To retain the predictive value of $P(x)$, we define mean or average of x without reference to experiments as

$$\mu = E(x) = \int_{-\infty}^{\infty} xP(x)dx. \tag{2.7}$$

Thus, μ is the expected mean or predicted mean by way of the function $P(x)$. μ is also called the theoretical mean. If an average is made of several results of measurement, one gets the experimental mean, $\bar{x}$, which should be a good estimate of the theoretical mean. In fact, in many situations, μ is unknown. We retain the symbol μ for the theoretical mean or expectation value and the symbol $\bar{x}$ or $\bar{y}$ for the sample mean, that is, the experimentally obtained averages of measurements of the random variables x, y and so on. From the theoretical mean, one can compute the variance, σ^2. The variance is the expectation value of the squared deviation of a random variable x from its theoretical mean μ. It is derived thus:

$$\sigma^2 = E[(x-\mu)^2] = \int_{-\infty}^{\infty}(x-\mu)^2 P(x)dx = \int_{-\infty}^{\infty} x^2 P(x)dx - \mu^2, \tag{2.8}$$

where the integral in Eq. 2.8 is to be understood as an improper integral, to be evaluated with limiting values. For completeness, the variance for a discrete random variable is given by:

$$\sigma^2 = \sum_{i=1}^{n} P(x_i) \cdot (x_i - \mu)^2 = \left(\sum_{i=1}^{n} P(x_i)x_i^2 \right) - \mu^2. \tag{2.9}$$

Note that the familiar expression for variance

$$\sigma^2 = \frac{1}{n}\sum_{i=1}^{n}(x_i - \mu)^2, \tag{2.10}$$

obtains when discrete weighted variance is specified by weights whose sum is not 1, that is, when the probabilities as we have defined already, are not used. The theoretical mean and the variance of various probability distribution functions $P(x)$ will be of interest in our derivations for the purposes of nuclear experiments.

Example 2.2 *Show that Eq. 2.10 reduces to Eq. 2.11 which is much easier to calculate in practise.*

$$\boxed{\sigma^2 = \frac{\sum_{i=1}^{n} x_i^2}{n} - \mu^2.} \tag{2.11}$$

Solution

We derive the variance beginning from its expanded form:

$$\sigma^2 = \frac{1}{n} \sum_{i=1}^{n} (x_i - \mu)(x_i - \mu) \tag{2.12}$$

$$= \frac{\sum_{i=1}^{n} (x_i - \mu)(x_i - \mu)}{n} \tag{2.13}$$

$$= \frac{\sum_{i=1}^{n} (x_i^2 - 2\mu x_i + \mu^2)}{n} \tag{2.14}$$

$$= \frac{\sum_{i=1}^{n} x_i^2}{n} - \frac{2\mu \sum_{i=1}^{n} x_i}{n} + \frac{\sum_{i=1}^{n} \mu^2}{n} \tag{2.15}$$

$$= \frac{\sum_{i=1}^{n} x_i^2}{n} - 2\mu\mu + \frac{n\mu^2}{n} \tag{2.16}$$

$$\doteq \frac{\sum_{i=1}^{n} x_i^2}{n} - \mu^2. \tag{2.17}$$

When samples have multiple variables, that is, instead of just the x_i (for example, various runs of gamma decay), we have x_i, y_i (for example various runs detecting gamma decay of different energies from a mixed sample containing two or more parent radioactive nuclei), the process will be characterized by a multivariate distribution $P(x,y,z,...)$. The theoretical mean and variance of each separate random variable retain the definitions given above, with integration over all variables. For instance, μ_y is the theoretical mean for y. In the probability density $P(x,y,z,...)$, however, there is a new important quantity, the *covariance*. The covariance is a measure of the linear correlation between pairs of variables in a multivariate distribution. For instance, in a trivariate distribution with $P(x,y,z)$, we have three covariances, $cov(x,y)$, $cov(x,z)$ and $cov(y,z)$. Of course, it is implied that the covariance is symmetric:

$$cov(x,y) = cov(y,x).$$

The theoretical covariance is given by:

$$cov(x,y) = E[(x - \mu_x)(y - \mu_y)]. \tag{2.18}$$

2.2 BINOMIAL DISTRIBUTION

The binomial distribution stems from the binomial theorem or binomial expansion, which states that it is possible to expand any powers n of $(x+y)^n$ into a sum. Consider a radioactive nucleus or a coin. If p is the probability of decaying or of being a head, and q is the probability of not decaying or of being a tail for the nucleus and coin respectively, then the probability of n decays or heads in N tries regardless of the order of occurrence is given by the binomial distribution:

$$P(n) = \frac{N!}{n!(N-n)!}p^n(1-p)^{N-n}. \tag{2.19}$$

The proof of Eq. 2.19 is given next as a guided exercise, in terms of the so-called Bernoulli process, that is, the events are independent so the sequence does not matter and the outcome remains only 2 possibilities: a success or a failure, a 1 or a 0.

Exercise 23

For n successes or decays (p) out of N trials, consider a sequence of outcomes $ppp...pqqq...q$. The ps occur n times and the qs occur $N-n$ times. The failures or survivors q are obviously $1-p$ for a Bernoulli process. Argue that

$$ppp...pqqq...q = p^n q^{N-n} = p^n(1-p)^{N-n}$$

because the outcome of consecutive experiments is independent. With this you have obtained the $p^n q^{N-n}$ part of Eq. 2.19. Now, argue that any sequence other than $ppp...pqqq...q$ containing the same number n of p out of N trials will still give $p^n(1-p)^{N-n}$. Finally, all possibilities here, without repetition, are given by the permutations *Perm* of N objects in sets of N when n of them are alike or have probability p, given by:

$$Perm(N,n) = \frac{N!}{n!(N-n)!} \tag{2.20}$$

Finally, multiply the Eq. 2.20 and $p^n(1-p)^{N-n}$ to get Eq. 2.19, which completes the proof.

Example 2.3 *Suppose the probability of success p in a binomial process is 0.2, what is the probability of getting 3 successes in 4 trials?*

Solution

Here, $N = 4$ and $n = 3$. Using Eq. 2.19,

$$P(0.2) = \frac{4!}{3!(4-3)!}0.2^3(1-0.2)^{4-3} = 0.0256.$$

Eq. 2.19 is discrete and can be summed up to obtain:

$$\sum_{n=0}^{n=N} P(n) = (p+q)^N. \tag{2.21}$$

For $p+q = 1$ (that is, the nucleus either decays or does not, the toss gives a head or a tail, so sum of probability is one), then, Eq. 2.21 is also one. Thus, $q = 1 - p$. This is the main idea behind binomial distribution, with emphasis on the *bi*, as either or. $P(n)$ is just the nth term of the binomial expansion, $(p+q)^N$. To check Eq. 2.19, we compute its average μ by evaluating the following sum:

$$\mu = \sum_{n=0}^{n=N} n\frac{N!}{n!(N-n)!}p^n(1-p)^{N-n} = Np. \tag{2.22}$$

The variance σ^2 of $P(n)$ can thus be calculated to obtain:

$$\sigma^2 = \sum_{n=0}^{n=N} (n-\mu)^2 P(n) = Np(1-p) = Npq. \tag{2.23}$$

Thus, the standard deviation for the binomial distribution is

$$\sigma = \sqrt{Npq}.$$

It gives an estimate of how much the results of measurements are expected to vary about the average value Np or μ.

Example 2.4 *Owing to vast applications in nuclear experiments [66], we consider an example with Cs-137 as the radioactive nucleus. Given that the half-life of Cs-137 is 30 years, what is the probability of decay for a nucleus of Cs-137, in one minute?*

Solution

The decay constant λ is actually the probability of one nucleus decaying per time.

$$\lambda_{Cs^{137}} = \frac{\ln 2}{(30y)(365d/y)(24h/d)(60min/h)} = 4.4 \times 10^{-8} \frac{1}{min}.$$

The probability p for decay is very small, just 4.4×10^{-8} per minute. Hence, $q = 1 - p \approx 1$ and so,

$$\sigma = \sqrt{Npq} \approx \sqrt{Np} \approx \sqrt{\mu}. \tag{2.24}$$

Note that in the above example, the relative probability for one nucleus to decay, p, for a time t, as opposed to not-decaying q based on the formulation of the binomial theorem is more formally given as

$$p = 1 - e^{-\lambda t}. \tag{2.25}$$

Obviously, the relative probability for one nucleus to not-decay, or to survive for a time t, is:

$$q = e^{-\lambda t}. \tag{2.26}$$

In fact, the radioactive decay law (chapter 1) can be derived using statistical reasoning as we now show in the next example.

Example 2.5 *From the empirical fact that radioactive decay is a random process with an average decay rate that characterizes a radionuclide, derive the radioactive decay law.*

Solution

For each radionuclide, the probability of decay p is directly proportional to the time interval of observation Δt, with a constant of proportionality λ, hence,

$$p = \lambda \Delta t. \tag{2.27}$$

Consistent with the Binomial theorem for a Bernoulli process, the probability of survival q during each Δt is: ·

$$q = 1 - p = 1 - \lambda \Delta t. \tag{2.28}$$

Thus, the probability of surviving 2 successive Δts is:

$$(1-p)(1-p) = (1-p)^2 = (1-\lambda\Delta t)^2.$$

For n successive intervals with duration t, such that $t = n\Delta t$,

$$q^n = (1-p)^n = (1-\lambda\Delta t)^n. \tag{2.29}$$

For this duration t,

$$\Delta t = \frac{t}{n},$$

hence, Eq. 2.29 becomes:

$$q^n = (1-p)^n = (1-\lambda\Delta t)^n = (1-\lambda\frac{t}{n})^n. \tag{2.30}$$

Noting that $\Delta t \to 0$ as $n \to \infty$, we apply the exponential expansion:

$$\lim_{n\to\infty}(1-\lambda\frac{t}{n})^n = e^{-\lambda t} = q. \tag{2.31}$$

Thus, we recover our expression for the probability of not decaying and so the number of atoms $N(t)$ that survive or remain un-decayed after a duration t is the radioactive decay law (Eq. 1.3):

$$N(t) = N_0 e^{-\lambda t}. \tag{2.32}$$

Note that when the probability of success or decay p is small (≤ 0.05), such that Np is finite, then the Binomial distribution can be approximated by the Poisson distribution which we present next.

2.3 POISSON DISTRIBUTION

In every day life, the number of photons hitting our retina (detector) within 10 minutes, the number of red cars going through an intersection, etc., can be predicted using the Poisson distribution. Simeon Denis Poisson (1781–1840) introduced the distribution along with his probability theory, in 1837[1], in his work with the title *Recherches sur la proba-bilite des jugements en matiere criminelle et en matiere civile* translated "Research on the Probability of Judgments in Criminal and Civil

---

[1]Though the result appeared earlier in a work by Abraham de Moivre, 1711, in De Mensura Sortis seu; de Probabilitate Eventuum in Ludis a Casu Fortuito Pendentibus.

Matters". Well, the connection with routine life and activities is well illustrated by our examples and Poisson's title. All counting events in high-energy physics and most of nuclear physics are regarded as Poisson-distributed[2] [76,77]. Yes, the number of decays in a given time interval in a radioactive sample follows a Poisson distribution [102]. The events are independent and have somehow a constant average rate of occurrence. The Poisson distribution is the limiting form of the binomial distribution when the probability $p \to 0$ and the total number of trials or events $N \to \infty$, even though the mean $\mu = Np$ stays finite. Using Eq. 2.19, the binomial distribution reduces to the Poisson distribution, for the probability of n events occurring:

$$P(n) = \frac{\mu^n e^{-\mu}}{n!}. \tag{2.33}$$

Like the binomial distribution, the Poisson distribution is discrete. The predictive power of the mean μ shows up nicely in the Poisson distribution, for even if we do not know Np, we can still find $P(n)$ using just μ. Recall that the conditions needed for $\mu \approx Np$ in the binomial distribution are actually the conditions satisfied in the Poisson distribution, namely, $p \to 0$, $N \to \infty$ and so $(1 - p) \to 1$. For this reason, radioactive decay processes are described by Poisson statistics since the random probability for a particular atom to decay per time p is very small. For instance, for Co-60 with a half-life of 5.26 years, p for a single Co-60 nucleus to decay in 1 s is

$$p = \lambda \Delta t = \lambda \times 1\mathrm{s} = \frac{0.693}{1.6588 \times 10^8} = 4.0 \times 10^{-9}.$$

Interestingly, if instead of the average number of events μ, we know the time rate λ for the events to happen then $\mu = \lambda t$ and

$$P(n) = \frac{(\lambda t)^n e^{-\lambda t}}{n!} = \frac{(\mu)^n e^{-\mu}}{n!}. \tag{2.34}$$

Note that μ must be greater than zero since the discrete random variables x of the Poisson distribution by definition are non-negative integers. Let us compute the variance σ^2 for the Poisson distribution using Eq.

---

[2]But keep in mind that formally, they are binomially distributed since the number of counts, of particles, etc., are still finite. The Poisson distribution relies on these numbers approaching infinity.

2.9 and Eq. 2.33:

$$\sigma^2 = \sum_{n=0}^{n=N} (n-\mu)^2 \frac{\mu^n e^{-\mu}}{n!} = \left(\sum_{n=0}^{\infty} n^2 \frac{\mu^n e^{-\mu}}{n!} \right) - \mu^2 = \mu. \qquad (2.35)$$

Now,

$$\left(\sum_{n=0}^{\infty} n^2 \frac{\mu^n e^{-\mu}}{n!} \right) - \mu^2 = \mu$$

because μ^2 is the square of the expectation value $E(x)$:

$$\mu^2 = (E(x))^2$$

and the term before it:

$$\left(\sum_{n=0}^{\infty} n^2 \frac{\mu^n e^{-\mu}}{n!} \right),$$

is the expectation value of the square of x, that is, $E(x^2)$ in discrete form with n. By definition,

$$\sigma^2 = E(x^2) - (E(x))^2.$$

To complete our computation,

$$\sum_{n=0}^{\infty} n^2 \cdot \frac{\mu^n e^{-\mu}}{n!} = \sum_{n=1}^{\infty} n^2 \cdot \frac{\mu^n e^{-\mu}}{n!} = \mu \sum_{n=1}^{\infty} n \cdot \frac{\mu^{n-1} e^{-\mu}}{(n-1)!} = \mu \sum_{j=0}^{\infty} (j+1) \cdot \frac{\mu^j e^{-\mu}}{j!}.$$

Multiplying out to extract another $E(x)$ term,

$$\mu \sum_{j=0}^{\infty} (j+1) \cdot \frac{\mu^j e^{-\mu}}{j!} = \mu \left(\sum_{j=0}^{\infty} (j) \cdot \frac{\mu^j e^{-\mu}}{j!} + \sum_{j=0}^{\infty} \frac{\mu^j e^{-\mu}}{j!} \right).$$

Inserting μ for the $E(x)$ term and rewriting the last term,

$$\mu \left(\sum_{j=0}^{\infty} (j) \cdot \frac{\mu^j e^{-\mu}}{j!} + \sum_{j=0}^{\infty} \frac{\mu^j e^{-\mu}}{j!} \right) = \mu \left(\mu + e^{-\mu} \sum_{j=0}^{\infty} \frac{\mu^j}{j!} \right) = \mu(\mu + e^{-\mu} e^{\mu}) = \mu(\mu+1).$$

Indeed,

$$\sigma^2 = \sum_{n=0}^{n=N} (n-\mu)^2 \frac{\mu^n e^{-\mu}}{n!} = \left(\sum_{n=0}^{\infty} n^2 \frac{\mu^n e^{-\mu}}{n!} \right) - \mu^2 = \mu(\mu+1) - \mu^2 = \mu.$$

Table 2.1 Poisson Distribution

n	μ	Poisson
0	2	0.135335283
1	2	0.270670566
2	2	0.270670566
3	2	0.180447044
4	2	0.090223522
5	2	0.036089409
6	2	0.012029803
7	2	0.003437087
8	2	0.000859272
9	2	0.000190949
10	2	3.81899E-05

Thus, for a Poisson-distributed random variable, we have the important result:

$$\boxed{\sigma^2 = \mu.}$$

(2.36)

This is a very useful result worth emphasizing: for any Poisson-distributed random variable, the variance σ^2 is equal to the mean μ. And obviously, the standard deviation is given by:

$$\sigma = \sqrt{\mu}.$$

(2.37)

Example 2.6 *Using values of x from 0 to 10, compute the Poisson distribution for $\mu = 2$, 3, 4 and 5 and show the results for $\mu = 2$ in a table. Plot the results for $\mu = 2$, 3, 4 and 5 in a figure.*

Solution

Using Eq. 2.33, the Poisson distribution is computed and Table 2.1 shows the result for $\mu = 2$. The required plots of Poisson distributions are shown in Fig. 2.1. Notice that as the mean increases, the distribution becomes more symmetric around the mean, illustrating the dependence of the distribution on the mean. In fact, we will show below that as the mean increases, the Poisson distribution approximates to the so-called

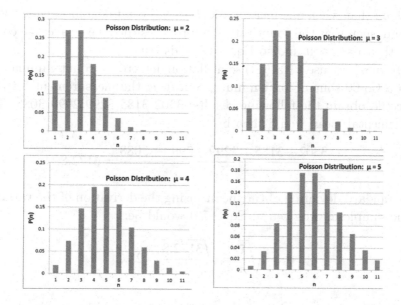

Figure 2.1 Plots of Poisson Distribution for values of the mean from 2 to 5. As the mean increases, the distribution becomes more symmetric around the mean.

normal or Gaussian distribution. More precisely, for $\mu \geq 20$, the Poisson distribution is indistinguishable from a Gaussian or normal distribution for the same sample and conditions.

Example 2.7 *In radiation "counting" measurements, it may not be feasible to repeat the measurements many times to obtain an average $\bar{x}$ and an uncertainty σ. In fact, common practice is to take one measurement x and state the result as $x \pm \sqrt{x}$. Justify this practice (a) theoretically and (b) experimentally.*

Solution

(a). Theoretically, radioactive decay is governed by Poisson statistics because it is inherently random and also the probability p of a single nucleus decaying per time, is very small. Using Eq. 2.38,

$$\sigma = \sqrt{\mu}, \qquad (2.38)$$

we claim that when one measurement x has been taken for a time t long enough, such that the count rate $\frac{n}{t}$ approaches the true or average count rate, then $x \approx \bar{x} \approx \mu$ and so Eq. 2.38 holds true.

(b). Assume, we use a NaI(Tl) scintillation detector to count gamma rays from a Co-60 source, for 1 minute, repeating the measurement 5 times. We could obtain the following results: $3302, 3185, 3299, 2990, 3088$. The experimental mean would then be

$$\bar{x} = \frac{3302 + 3185 + 3299 + 2990 + 3088}{5} = 3172.8.$$

The standard deviation of the mean (using the derivation of the variance of the sample mean given in Eq. 2.73) would be:

$$\sigma_{\bar{x}} = \sqrt{\frac{\bar{x}}{n}} = \sqrt{\frac{3172.8}{5}} = 25.2.$$

The fractional standard deviation of the mean would be obtained as:

$$\frac{\sigma_{\bar{x}}}{\bar{x}} = \frac{25.2}{3172.8} = 0.0079$$

But because $\sigma = \sqrt{x}$ for each independent count,

$$\sigma_{total} = \sqrt{total}$$

and therefore, assuming the experimental mean is close to the true mean, the Poisson-predicted standard deviation would be:

$$\sigma = \sqrt{\bar{x}} = \sqrt{3172.8} = 56.3$$

The fractional uncertainly here is therefore

$$\frac{\sigma}{\bar{x}} = \frac{56.3}{3172.8} = 0.0178.$$

If, instead of 5 repeats, we measure for 5 minutes, we could obtain 15864 counts. The uncertainty obtained in the 5 minute count, based on Poisson statistics, is:

$$\sigma = \sqrt{15864} = 125.95.$$

Hence, the fractional uncertainty becomes:

$$\frac{\sigma}{\bar{x}} = \frac{125.95}{15864} = 0.0079.$$

The smaller fractional uncertainty with a single long measurement is the reason that we seek to improve statistical precision of radioactive measurements by counting for longer times. By precision, we mean the reproducibility and not the accuracy. Numerically, statistical precision is obtained using the fractional uncertainty or fractional standard deviation which is also termed the *coefficient of variation*, CV, often expressed in percentage:

$$\%CV = \frac{\sigma}{\bar{x}} \times 100. \tag{2.39}$$

Note that the inverse of the *CV* gives the signal to noise ratio (SNR):

$$SNR = \frac{\bar{x}}{\sigma}. \tag{2.40}$$

For, Poisson distribution,

$$\%CV = \frac{\sigma}{\mu} \times 100 = \frac{\sqrt{\mu}}{\mu} \times 100 \tag{2.41}$$

Hence, the precision or %CV of the 5-minute count illustrated above was about 0.8% while that of the 1-minute count was about 2%.

Important note: It is tempting to claim the square root of a derived quantity involving counts as the standard deviation of that quantity. Since this mistake is somehow common, the warnings against it are also strong. Here is one strong warning from the literature [61]: "In radioactive decay or nuclear counting, we may directly apply $\sigma = \sqrt{x}$ *only* if x represents.... a number of events over a given observation time recorded from a detector. The vast majority of mistakes made in the use of counting statistics results from the misapplication of above relation. ....One *cannot* associate the standard deviation σ with the square root of any quantity that is not a directly measured number of counts. For example, the association does *not* apply to:

1. Counting *rates*

2. *Sums* or *differences* of counts

3. *Averages* of independent counts

4. Any *derived* quantity"

Example 2.8 *After using a scintillating detector to measure the count rate r(t) of a radioactive sample for 10 minutes, which you found to be 2300 counts per minute (cpm) you now wish to estimate the standard deviation or uncertainty σ associated with this measurement. Since radioactivity is a Poisson process, you are tempted to apply the formula*

$$\sigma = \sqrt{\mu}.$$

(a) Use dimensional analysis to show that this is wrong and note that it is also wrong for the fact that a derived quantity (here, count rate) is not a Poisson process just because it contains a variable that is a Poisson process. (b) Obtain a correct expression for the calculation of the uncertainty in the measurement.

Solution

(a) We know that count rate $r(t)$ is given by

$$r(t) = \frac{n}{t}$$

where n is number of counts and t is time. Now, if one uses

$$\sigma = \sqrt{\mu} = \sqrt{r(t)},$$

then, the result will have to be stated as $r(t) \pm \sqrt{r(t)}$ and this is dimensionally inconsistent. That is, $r(t)$ has unit s^{-1} while its estimated error is now given with unit of $\sqrt{s^{-1}}$, which is profoundly wrong. (b) Since the radioactive count n is a Poisson variable,

$$\sigma = \sqrt{\mu} = \sqrt{n} = \sqrt{r(t) \times t}$$

and

$$r(t) = \frac{n}{t}$$

we have a change of scale situation and the *change of scale rule of standard deviation* is applied to give:

$$\sigma_{r(t)} = \frac{\sqrt{n}}{t} = \frac{\sqrt{r(t) \times t}}{t} = \frac{\sqrt{r(t) \times t}}{\sqrt{t} \times \sqrt{t}} = \sqrt{\frac{r(t)}{t}} \qquad (2.42)$$

Thus, the result will simply be stated as

$$r(t) \pm \frac{\sqrt{r(t) \times t}}{t} = r(t) \pm \sqrt{\frac{r(t)}{t}}$$

or more transparently,

$$\frac{n}{t} \pm \frac{\sqrt{n}}{t}.$$

Of course, when time is not considered a constant as in this case but a variable with its own uncertainty, then full error propagation formulas (treated later in this chapter) are applied. In most cases, the uncertainty in the measurement of time is very small and so treating it as a constant is well justified.

Example 2.9 *You have a radioactive sample with an activity of 37 decays per second (37 Bq). Taking this activity as the mean decay rate, what is the probability of counting exactly 5 transformations in 1 s?*

Solution

Using the Poisson distribution (Eq. 2.34),

$$P(5) = \frac{(\mu)^n e^{-\mu}}{n!} = \frac{(37)^5 e^{-37}}{5!} = 4.93 \times 10^{-11}$$

Thus, getting 5 decays in 1 s is very unlikely since the average counts in 1 s is theoretically 37 in this case.

Example 2.10 *For the radioactive sample with an activity of 37 decays per second (37 Bq) what is the probability of counting exactly 38 transformations in 1 s?*

Solution

Again, using the Poisson distribution (Eq. 2.34),

$$P(38) = \frac{(37)^{38} e^{-37}}{38!}.$$

Although online calculators and certain other calculators can handle 38!, some may not. In that case, any of the many versions of the so-called Stirling's approximation (first used by Abraham de Moivre) can be used:

$$\boxed{n! \approx \sqrt{2\pi n} \times (\frac{n}{e})^n} \qquad (2.43)$$

So

$$P(38) = \frac{(37)^{38}e^{-37}}{38!} = 0.0637.$$

Thus, the probability of obtaining 38 counts in 1 s when the average rate is 37 is much higher than that of getting 5 in 1 s. The Poisson distribution gives the range of observations around the theoretical mean of 37 dps in this case. The width of the range of observations is given by σ, the standard deviation.

Exercise 24

Show that the fractional standard deviation or coefficient of variation for the Poisson distribution is

$$CV = \frac{1}{\sigma} = \frac{1}{\sqrt{\mu}}.$$

Hint: begin from

$$\sigma_{Poisson} = \sqrt{\mu} = \sqrt{\bar{x}}$$

and

$$CV = \frac{\sigma}{\bar{x}}$$

2.4   GAUSSIAN OR NORMAL DISTRIBUTION

The probability density function for the normal distribution, also known as the Gaussian distribution which is a continuous distribution of the random variable x is given as [55]:

$$P(x) = \frac{1}{\sqrt{2\pi\sigma^2}}e^{-\frac{(x-\mu)^2}{2\sigma^2}} \qquad (2.44)$$

where the mean μ and variance σ^2 retain the definitions given in Eq. 2.7 and Eq. 2.8 respectively. The mean is given by:

$$\mu = \int_{-\infty}^{\infty} x\frac{1}{\sqrt{2\pi\sigma^2}}e^{-\frac{(x-\mu)^2}{2\sigma^2}}\,dx. \qquad (2.45)$$

Thus, the normal distribution is symmetrical about $x = \mu$. Using Eq. 2.8, the variance of the Gaussian distribution is given by:

$$\sigma^2 = \int_{-\infty}^{\infty} x^2\frac{1}{\sqrt{2\pi\sigma^2}}e^{-\frac{(x-\mu)^2}{2\sigma^2}}\,dx - \mu^2. \qquad (2.46)$$

Under what conditions can the binomial distribution be approximated by the normal distribution and under what conditions can the same binomial distribution be approximated by the Poisson distribution? When n is large or approaches infinity while p and $1 - p$ (see Eq. 2.19), are not too small, then the binomial distribution can be approximated by the normal distribution with $\mu = np$ and $\sigma^2 = np(1 - p)$. In practice, this approximation works well when both np and $n(1-p)$ exceed 4 [55]. A simple visual inspection shows that for $\mu = np = 5$ in Fig. 2.1, one sees already a normal distribution emerging, instead of a typical unsymmetric Poisson distribution with $\mu = 2$, for instance. The shape of the normal distribution is instructive. The standard deviation σ shows up as a measure of the width of the distribution. The very practical measure, full width at half maximum, FWHM, is just

$$FWHM = \sigma\sqrt{8\ln 2} = 2\sigma\sqrt{(2\ln 2)} = 2.35\sigma. \qquad (2.47)$$

It is standard practice in nuclear experiments to use the Full-Energy Width at Half the Maximum peak (FWHM) above the background as the measure of the photopeak width, in order to obtain the experimental standard deviation. This is because the FWHM is proportional to the standard deviation of the distribution's mean value, given by the centroid [91]. Gaussian fits to the counts versus energy data as in Fig. 2.2 enable reliable determination of FWHM.

Example 2.11 *From the 511 keV annihilation photon peak of Fig. 2.3, you obtain FWHM of 46.52 keV. (a) What is the standard deviation σ for this peak? (b) The centroid of this 511 keV peak is 509.37 keV. Is this centroid reasonably close to the expectation value of 511 keV?*

Solution

(a) Using Eq. 2.47, $\sigma = \frac{FWHM}{2.35} = 19.80\,\text{keV}$.
(b) The centroid is reasonaby close: it lies within 1 σ of the expectation value or mean, namely, 511 keV. Hence, the result could be presented as 509 ± 19 keV. Obviously, the narrower the Gaussian peak, the smaller the σ.

Example 2.12 *You are preparing for a NIM experiment in which you wish to obtain an error limit of 0.1% with a confidence interval of 95%. How many counts should you record?*

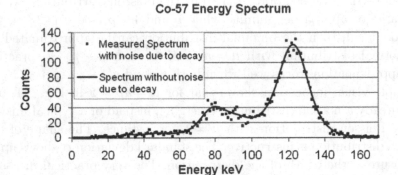

Figure 2.2 Counts versus Energy Spectrum of Co-57 showing random errors around the solid line due to random nature of radioactive decay itself. Image adapted from Nuclear Medicine Physics: A Handbook for Teachers, p. 151. IAEA 2014

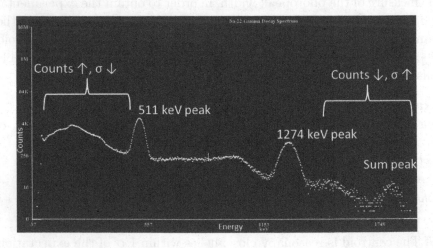

Figure 2.3 Counts versus Energy Spectrum of Na-22 showing less random errors or lower standard deviation σ (less spread) at regions with higher counts and more random errors or higher standard deviation σ at regions with lower counts. Data was taken by the author using a 5.08 cm $\times$ 5.08 cm NaI(Tl) detector and recorded using the UCS 30 multichannel analyser and software (Spectrum Techniques).

Solution

For 95% confidence interval, the counts should fall within 2σ or 2 standard deviations (see Table 2.3). Thus

$$0.1\% \; error \; limit = \frac{2\sigma}{N} = 2\frac{\sqrt{N}}{N},$$

$$0.001 = 2\frac{\sqrt{N}}{N} = \frac{2}{\sqrt{N}}$$

$$(0.001)^2 = \frac{4}{N}.$$

Hence, $N = 4 \times 10^6$ counts.

Example 2.13 *In a NIM experiment, you wish to ensure a statistical uncertainty that is not more than 3%. What minimum number of counts do you need to detect?*

Solution

Here the relative uncertainty is given as a percentage and radioactive decay counting is a Poisson process, hence,

$$3\% \; error \; limit = \frac{\sigma}{N} = \frac{\sqrt{N}}{N} = \frac{1}{\sqrt{N}},$$

$$0.03 = \frac{\sqrt{N}}{N} = \frac{1}{\sqrt{N}}.$$

Therefore,

$$N = \left(\frac{1}{0.03}\right)^2 = 1111 \; counts$$

Exercise 25

From the 1274 keV gamma peak of Fig. 2.3, you obtain FWHM of 74.44 keV. (a) What is the standard deviation σ for this peak? (b) The centroid of this 1274 keV peak is 1277.72 keV. Is this centroid reasonably close to the expectation value of 1274 keV?

2.5 MAXWELLIAN DISTRIBUTION

Nuclear fission and fusion can lead to the production of free neutrons. Free neutrons are unstable: they decay into a proton, an electron, an anti-electron-neutrino with a mean lifetime τ of about 887 seconds. The speed and energy distribution of thermal neutrons (slow neutrons) follow the Maxwell distribution which is also the energy distribution of a gas in thermodynamic equilibrium. Using kinetic energy E and speed v as random variables, with

$$E = \frac{1}{2}m_n v^2,$$

$$P(E) = \frac{2\pi P_0}{(\pi KT)^{3/2}}\sqrt{E}e^{-\frac{E}{KT}} \qquad (2.48)$$

where K is the Boltzmann's constant; T is the absolute temperature of the medium and P_0 is the total number of neutrons per unit volume,

$$P_0 = \int_0^\infty N(v)dV = \int_0^\infty N(E)dE.$$

Likewise,

$$P(v) = \frac{4\pi v^2 P_0}{(2\pi KT/m)^{3/2}}e^{-\frac{mv^2}{2KT}}. \qquad (2.49)$$

2.6 CHI-SQUARE DISTRIBUTION

The Chi-Square distribution χ^2 is an important and versatile continuous probability distribution whose origin and development have been traced back to Bienayme in 1838, Abbe in 1863, Maxwell in 1860 and Boltzmann in 1881 [51]. The Chi-Square distribution χ^2 is used to test whether an experimental frequency distribution (from samples) is consistent with a theoretical distribution. This usage as a statistical test procedure came through the seminal paper of Karl Pearson in 1900 [89], hence, the so-called Pearson's chi-square test. In more common terms, it is used to test the *goodness-of-fit* of theoretical formulas to experimental data. In nuclear physics experiments, the theoretical distributions are usually the Poisson and Gaussian distributions. Thus, in Nuclear Instruments and Methods and allied fields involving radioactivity, the χ^2 test enables assessment of precision in counting instruments [26, 125] and whether random variations in a set of measurements are consistent with what should be expected for a Poisson distribution or a normal or Gaussian distribution. If the measurement of a random variable x has been repeated N times giving data-set x_i, then the χ^2 is defined as:

$$\chi^2 = \sum_{i=1}^{N} \frac{(x_i - \bar{x})^2}{\sigma_i^2}, \qquad (2.50)$$

where σ_i are the variances for each x_i and $\bar{x}$ is the sample or arithmetic mean of all the measurements. Note that the sample mean $\bar{x}$ can be calculated as follows (as we will show in the next section, see Eq. 2.68):

$$\bar{x} = \frac{1}{n} \sum_{i=1}^{n} x_i.$$

A cursory look at Eq. 2.50 reveals that χ^2 can also be defined using only the x_i and their average:

$$\chi^2 = \frac{1}{\bar{x}} \sum_{i=1}^{N} (x_i - \bar{x})^2. \qquad (2.51)$$

From these definitions, one sees that the Chi-Square distribution is not symmetric. Furthermore, the numerical value of χ^2 depends on number of measurements N, which is often designated the number of degrees of freedom DF. More formally, when considered as the number of values in a set of measurements that can be changed without significantly changing the value of the sum,

$$DF = N - 1$$

Intuitively, DF is the number of ways the observed distribution may differ from the theoretically expected distribution. DF depends on the expected theoretical distribution function. It can be $N - 2$, $N - 3$, depending on number of fixed parameters involved in calculating the theoretical distribution function. Thus, for Poisson distribution,

$$DF_{Poisson} = N - 2$$

and for the Gaussian distribution,

$$DF_{Gaussian} = N - 3$$

Once, the DF is found, then the probability $P(\chi^2, DF)$ that the observed deviations agree with the theoretically expected distribution function is calculated, using Chi Square Tables composed from Eq. 2.52 for $P(\chi^2, DF)$ [51]:

$$P(\chi^2, DF) = \begin{cases} \frac{\chi^{(\frac{DF}{2}-1)}e^{-\chi/2}}{\Gamma(\frac{DF}{2})2^{DF/2}} & \text{if } \chi > 0, \\ 0 & \text{if } \chi \leq 0, \end{cases} \tag{2.52}$$

where Γ is the Gamma function, which is the extension of the factorial to both real and complex number arguments, given by:

$$\Gamma(DF) = (DF - 1)! \tag{2.53}$$

and calculated for restricted cases using its definite integral form for $|x| > 0$:

$$\Gamma(x) = \int_0^\infty t^{x-1}e^{-t}dt. \tag{2.54}$$

Eq. 2.52 is the probability density function for the χ^2 distribution (Eq. 2.50). Fig. 2.4 shows plots of $P(\chi^2, DF)$, which correspond to the probability of exceeding χ^2 for a given degree of freedom, based on demonstration plots from Mathematica [124]. Tables of values of $P(\chi^2, DF)$ can be obtained from texts or online sources such as MedCalc, an easy-to-use statistical software (where the table is given for DF up to 1000: https://www.medcalc.org/manual/chi-square-table.php). In consulting tables or calculating $P(\chi^2, DF)$ for oneself, care should be taken to clarify whether the P-values used or given are for the probability of being less than χ^2 or are for exceeding χ^2 just as we have indicated in Fig. 2.4. Linear interpolations are often needed to get P for values of χ^2 lying between two columns of P. When $P(\chi^2, DF) = 0.5$, it means that there is a 50% probability of getting a more spread out distribution of measurements and a 50% probability of getting a narrower distribution if the series of measurements is repeated. A very important use of χ^2 is for determining the probability that an observed departure from the expected distribution is due to random fluctuations or whether it is due to systematic errors such as equipment malfunctioning. Within 95% confidence interval ($0.025 < \chi^2 < 0.975$), χ^2 value is considered a satisfactory indication that the measurement system is without systematic errors. In other words, fluctuations are only due to the random or statistical nature of the process measured. Furthermore, to quantify the "goodness"

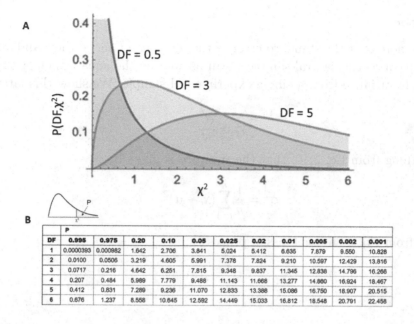

Figure 2.4 A. Plots of $P(\chi^2, DF)$, corresponding to the probability of exceeding χ^2 for a given degree of freedom, DF, based on demonstration plots from Mathematica [124]. B. A typical table of values of $P(\chi^2, DF)$ MedCalc, an easy-to-use statistical software (where the table is given for DF up to 1000: https://www.medcalc.org/manual/ chi-square-table.php). There are caveats for tables such as this. In some, P-values are given for the probability of being less than χ^2 for a given DF.

of fit or the "normality" of the data using the the value of P, a common choice of range of acceptability is to set P at 0.05 and 0.95. Any data set for which P falls into this range is considered "good" within a 90% confidence level for the goodness of the data.

Example 2.14 *Show that for Poisson processes,*

$$\chi^2 = \frac{N\sigma^2}{\bar{x}}. \tag{2.55}$$

Solution

Here, note that the standard error of the mean can refer to the standard deviation of the several-sample mean as well as the square-root of the sample variance from a single experimental sample. We show this later in Eq. 2.73.

$$s_{\bar{x}}^2 = \frac{\sigma^2}{N},$$

Recalling from Eq. 2.10, that the variance is given by,

$$\sigma^2 = \frac{1}{n}\sum_{i=1}^{n}(x_i - \mu)^2,$$

and from Eq. 2.36, for Poisson processes,

$$\sigma^2 = \mu,$$

hence,

$$\sigma_i^2 \approx \bar{x}.$$

Thus, Eq. 2.50 becomes

$$\chi^2 = \sum_{i=1}^{N}\frac{(x_i - \bar{x})^2}{\sigma_i^2} = \frac{N\sigma^2}{\sigma_i^2} = \frac{N\sigma^2}{\bar{x}}. \tag{2.56}$$

In the section on "Determination and Tests of Probability Distributions", we will illustrate the use of Chi-Square tests to determine the goodness-of-fit or the extent to which experimental data follow theoretically predicted probability distributions.

2.7 EXPONENTIAL DISTRIBUTION

In general, we notice many exponential functions in the distributions that are relevant for nuclear physics. Even the exponential decay function found in the Radioactive Decay Law, shares properties of the following exponential density distribution function:

$$\boxed{P(x) = \lambda e^{-\lambda x}.} \tag{2.57}$$

$$\mu = \frac{1}{\lambda}, \tag{2.58}$$

$$\sigma^2 = \int_0^\infty x^2 \lambda e^{-\lambda x} - \mu^2 = \frac{1}{\lambda^2}. \tag{2.59}$$

Thus for any exponentially distributed random variable,

$$\boxed{\sigma^2 = \mu^2.} \tag{2.60}$$

A good example of an exponentially distributed function in nuclear experiments is the proper decay time t of an unstable particle or radioisotope.

2.8 LANDAU AND OTHER DISTRIBUTIONS

Other probability distributions appear in nuclear physics and are used for making sense of experimental data [5, 73]. We list a few of them, and adumbrate on their fundamental mathematical structure as well as their physical utility in nuclear data analysis.

1. Power Law Probability Distribution

2. Log-normal Probability Distribution

3. Landau Distribution

4. Cauchy Distribution

5. Gamma Distribution

6. Uniform Distribution

The *power law probability distribution* emerges from microscopic theories and help relate them to observable macroscopic phenomena. Derivation of the power law distribution entails that certain experimental regimes are scale free, that is, they do not possess a preferred scale in space, time, or something else [113]. The *lognormal distribution function* is used to represent inherently positive physical random quantities. Instead of a formal definition, here is the connection between the normal or Gaussian distribution with the lognormal distribution. When the probability function for a random variable x is assumed to be lognormal, then the distribution for $y = lnx$ is the normal distribution. Since the converse is also true, the normal distribution and the lognormal are each other's converse. The *Landau distribution* characterizes fluctuations in energy losses by charged particles as they traverse matter. It was derived

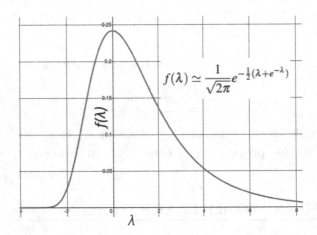

$$f(\lambda) \simeq \frac{1}{\sqrt{2\pi}} e^{-\frac{1}{2}(\lambda + e^{-\lambda})}$$

Figure 2.5 The Landau Distribution plotted based on its approximate functional form.

by Lev Landau. Since the Bethe-Bloch formula [11, 17] (see chapter 3) describes only the average energy loss of charged particles traversing matter, the Landau distribution is important as it provides further details about the process of energy losses. It provides a universal asymmetric probability density function which has a narrow peak with a long tail for positive values. The tail comes from the small number of individual collisions, each with a small probability of transferring comparatively large amounts of energy. The Landau probability density function can be stated thus:

$$f(\lambda) = \frac{1}{2\pi i} \int_{c-i\infty}^{c+\infty} e^{s\ln s + \lambda s} ds, \quad c \geq 0 \tag{2.61}$$

which is in the form of a Laplace transformation. In practical applications, the Landau distribution is given approximately by:

$$f(\lambda) \simeq \frac{1}{\sqrt{2\pi}} e^{-\frac{1}{2}(\lambda + e^{-\lambda})}, \tag{2.62}$$

where λ parameterizes the difference between the actual energy loss and the average energy loss. Figure 2.5 shows a typical Landau distribution for ionizing particles traversing through matter. The detection efficiency of some nuclear detectors such as the high purity germanium detector, follows a Landau efficiency distribution. The *Cauchy Distribution*, also called *Lorentz Distribution*, *Breit–Wigner Distribution* or any combination of these eponyms, can be used as an alternative to the Gaussian distribution in modeling random noise in nuclear experiments or statistical noise in nuclear data, neutron lifetimes [99], etc. It has longer tails than the Gaussian distribution. You will not question the categorisation of the Cauchy distribution as a "pathological function" when you discover that its expectation value and variance are undetermined! However, fractional moments do exist for the Cauchy distribution. Its probability density function is:

$$P(x) = \frac{1}{\pi}\left[\frac{\gamma^2}{(x-x_0)^2 + \gamma^2}\right], \tag{2.63}$$

where γ is the scale parameter specifying the half-width at half maximum (HWHM), x_0 is the location parameter. Note that when the scale parameter is infinitessimal, we obtain the Dirac delta function. The standard Cauchy distribution is obtained for the special case of $\gamma = 1$ and $x_0 = 0$:

$$P(x,0,1) = \frac{1}{\pi(1+x^2}\tag{2.64}$$

The *Gamma Distribution* has been somewhat treated already because the exponential distribution and the Chi-Square distribution are in fact special cases of the gamma distribution. In general, two variables are enough to parameterize the gamma distribution. We see these clearly in the gamma function explicitly given in Eq. 2.53.

The *Uniform Probability Distribution* (sometimes called rectangular distribution, depending on dimensions) has boundaries or limits which are usually specified as $\pm b$ from the central value, where b is the half-width of the distribution [33]. Hence, 100% of the values fall between $-b$ and $+b$. The uniform probability distribution provides the *least* information since all that is known are the limits within which a value will fall. It $-b$ is stated more generally as just a, then the mean μ of the uniform distribution is just:

$$\mu = \frac{a+b}{2}\tag{2.65}$$

and the variance σ^2 is:

$$\sigma^2 = \frac{(b-a)^2}{12}.\tag{2.66}$$

Hence, the standard deviation σ of a rectangular distribution when the limits are $\pm b$ is given by:

$$\sigma = \frac{b}{\sqrt{3}}.\tag{2.67}$$

We often make assumptions that the distribution of quantities in the nucleus such as charges is uniform, thus, the uniform distribution models situations where the probability of obtaining any value between two stated limits is equal to the probability of obtaining any other value between these limits. It can be applied to many dimensions and geometries such as 3D (dimensionally) and cubes or spheres.

2.9 DETERMINATION AND TESTS OF PROBABILITY DISTRIBUTIONS

When the true probability density function $P(x)$ is unknown as does happen following experiments, then from a finite sample of n events, a probability distribution function may be extracted with the assumption that there is such an underlying function. From such a finite-sized sample of n events, parameters that characterize the probability distribution can be extracted. Sometimes, this procedure of determination of distribution function and extraction of parameters is termed *shape characterization*. It also involves so-called goodness-of-fit tests. Major steps include estimation of mean, variance and Chi-Square Tests. We will illustrate these steps extensively in this section since a major aspect of Nuclear Physics experiments beyond data taking is the determination and test of probability distributions for data characterization.

2.9.1 Estimation of Mean, Variance and Covariance from Samples

If we take a sample of size n for a variable x, that is, $x_i = x_1, x_2, x_3 ... x_n$ from a distribution whose theoretical mean and variance are μ and σ^2 respectively, but we do not know the functional form of the distribution, then the *sample mean* or *experimental mean, $\bar{x}$* can be used to estimate μ and σ^2. The sample mean $\bar{x}$ is calculated from the data as follows:

$$\bar{x} = \frac{1}{n} \sum_{i=1}^{n} x_i. \tag{2.68}$$

The sample mean $\bar{x}$ is used as a good (in fact best) estimate of the theoretical mean μ of the sought-after distribution. This is because the sample or empirical mean $\bar{x}$ is an unbiased and convergent estimation of the theoretical mean, μ for in the limit $n \to \infty$, $\bar{x} \to \mu$:

$$\mu = \lim_{x \to \infty} \frac{1}{n} \sum_{i=1}^{n} x_i. \tag{2.69}$$

Eq. 2.69 can further be demonstrated using the expectation value of the mean:

$$E(\bar{x}) = \frac{1}{n} \sum_{i=1}^{n} E(x).$$ (2.70)

Likewise, the theoretical variance σ^2 can be estimated using the sample mean-squared deviations or sample variance which we denote as s^2:

$$s^2 = (\bar{x^2}) - (\bar{x})^2,$$ (2.71)

or more explicitly:

$$\boxed{s^2 = \frac{1}{n} \sum_{i=1}^{n} (x_i - \bar{x})^2.}$$ (2.72)

In the limit $n \to \infty$, $s^2 \to \sigma^2$. However, for a finite-sized sample, the sample variance s^2 systematically underestimates the theoretical variance, σ^2. This can be demonstrated by considering both the sample variance and the variance of the experimental or sample mean (not just the variance of population or theoretical mean), $s_{\bar{x}}^2$ and their expectation values:

$$s_{\bar{x}}^2 = E[(\bar{x} - \mu)^2] = \frac{\sigma^2}{N},$$ (2.73)

and so the expectation value of the sample variance becomes:

$$E(s^2) = \sigma^2 - s_{\bar{x}}^2 = \frac{n-1}{n} \sigma^2.$$ (2.74)

To overcome this systematic underestimation, a modified experimental variance is usually used and we can now see the mathematical origin of its familiar form:

$$s^2 = \frac{1}{n-1} \sum_{i=1}^{n} (x_i - \bar{x})^2.$$ (2.75)

For finite sized samples, Eq. 2.75 ensures an unbiased estimation of the variance σ^2, in which case, the sample mean $\bar{x}$ serves as estimate of the theoretical mean μ. Intuitively, the $\frac{1}{n}$ factor in Eq. 2.68 implies that the experimental average can be estimated even on the smallest sample consisting of just one event. Similarly, the $\frac{1}{n-1}$ factor in Eq. 2.75 implies that at least two events are needed to estimate their experimental variance or dispersion.

Just as the sample or experimental variance s^2 can be calculated from the sample mean, the sample or experimental covariance $cov_{exp}(x,y)$ is obtained thus:

$$cov_{exp}(x,y) = \frac{1}{n}\sum_{i=1}^{n-1}(x-\bar{x})(y-\bar{y}). \tag{2.76}$$

2.9.2 Using Relative Frequency for Sample Mean and Variance

In practice, experimental data sets are conveniently represented using a *relative frequency distribution*, $F(x)$, from which can be calculated all the aforementioned sample statistical moments: sample mean $\bar{x}$, sample variance s^2, etc. By definition,

$$\text{Relative Frequency } F(x) \equiv \frac{n_x}{N} \tag{2.77}$$

where n_x is the number of occurrences of the value x and N is the total number of measurements. Clearly, $F(x)$ is normalized:

$$\sum_{x=0}^{\infty} F(x) = 1. \tag{2.78}$$

The experimental or sample mean $\bar{x}$ can therefore be calculated using Eq. 2.77:

$$\bar{x} = \sum_{x=0}^{\infty} xF(x). \tag{2.79}$$

Example 2.15 *Table 2.2 shows the relative frequencies $F(x)$ for a random variable x. What is the sample mean $\bar{x}$?*

Table 2.2 Relative frequencies for a random variable

x	1	3	4	5	7
F(x)	0.3	0.15	0.15	0.2	0.2

Solution

Using Eq. 2.79,

$$\bar{x} = \sum_{x=0}^{\infty} xF(x) = (1 \times 0.3) + (3 \times 0.15) + (4 \times 0.15) + (5 \times 0.2) + (7 \times 0.2)$$

$$= 3.75$$

Using the sample mean, the amount by which any data point differs from sample mean, termed *residual*, can be calculated:

$$\text{residuals} \equiv x_i - \bar{x}. \tag{2.80}$$

The residuals sum up to zero, hence, the square of the residuals shows up in the modified sample variance, analogous to the square of the deviations in the sample variance and so previous equations hold, namely, Eq. 2.72 for sample variance and Eq. 2.75. Consistent with Eq. 2.72, the sample variance s^2 is then obtained from the relative frequency thus:

$$\boxed{s^2 = \sum_{x=0}^{\infty} (x_i - \bar{x})^2 F(x).} \tag{2.81}$$

2.9.3 Chi-Square Test and other Statistical Tests on Experimental Data

In defining and describing the Chi-Square distribution χ^2 above, we stated its application in testing whether an experimental frequency distribution (from samples) is consistent with a theoretical distribution. To illustrate this application, we use the χ^2 test on several runs of radioactive decay measurements, to measure how consistent the frequency distribution of each of the runs is with two theoretically expected distributions, namely, the Poisson and the Gaussian distributions. Using the definitions of the χ^2 distribution in Eq. 2.50 and Eq. 2.51 and considering the residuals between experimental and theoretically predicted values (see Eq. 2.80), the χ^2 value for testing is obtained thus:

$$\chi^2 = \sum_{i=1}^{N} \frac{(Residual_i)^2}{Theoretical_i}$$

$$= \sum_{i=1}^{N} \frac{(observed\,frequency - theoretical\,frequency)^2}{theoretical\,frequency}. \tag{2.82}$$

For instance, when finding the χ^2 for an experimental run vis-a-vis a Poisson distribution, Eq. 2.82 takes the form of Eq. 2.83 that we showed earlier:

$$\chi^2 = \sum_{i=1}^{N} \frac{(x_i - \bar{x})^2}{\sigma_i^2} = \frac{N\sigma^2}{\sigma_i^2} = \frac{N\sigma^2}{\bar{x}}. \qquad (2.83)$$

Here are the steps to follow in doing a typical χ^2 test in the context of nuclear physics experiments and nuclear medicine practise.

1. Compute sample mean $\bar{x}$ using Eq. 2.79,

$$\bar{x} = \sum_{x=0}^{\infty} xF(x)$$

   or Eq. 2.68

$$\bar{x} = \frac{1}{n}\sum_{i=1}^{n} x_i.$$

2. Compute χ^2 using Eq. 2.50 or Eq. 2.82 for the distribution being tested (eg, Poisson or Gaussian).

3. Determine degrees of freedom DF based on total number of measurements N.

4. Apply a *rule of thumb* test by finding $\frac{\chi^2}{DF}$. Eg, for Poisson distribution, it should be approximately 1 since Eq. 2.83 has $\frac{\sigma^2}{\bar{x}}$ approaching 1 as the experimental distribution approaches the Poisson distribution. Thus, for Poisson and Gaussian distributions, $\frac{\chi^2}{DF}$ should range from 0.5 to 2 otherwise, the experimental distribution is considered non-Poisson and non-Gaussian, that is:

$$0.5 \leq \frac{\chi^2}{DF} \leq 2$$

5. Perform a rigorous test by calculating the probability $P(\chi^2, DF)$ that the observed deviations agree with the theoretically expected distribution function, using Tables composed based on Eq. 2.52. Set confidence levels for P and evaluate. For instance, for a 90% confidence level, limits for P are:

$$0.05 \leq P \leq 0.95$$

A

Run	Dwell Time (s)	Mean	Variance	Standard Deviation	Standard Error	Poisson Estimate σ_P	$\sigma_{<n>}$	$\frac{\Delta\sigma}{\sigma}$
1	0.02	1.045	0.963	0.981	0.069	1.022	0.072	0.0418
2	0.04	2.230	2.177	1.475	0.104	1.493	0.106	0.0122
3	0.1	5.175	4.974	2.230	0.158	2.275	0.161	0.0202
4	0.8	41.700	39.700	6.301	0.446	6.458	0.457	0.0249
5	1.0	54.205	56.473	7.515	0.531	7.362	0.521	0.0204
6	2.0	106.36	97.970	9.898	0.700	10.313	0.729	0.0419

B

Run	Dwell Time (s)	Mean Value	Mean χ^2	P	Poisson χ^2	DF	χ^2/DF	P	Gaussian χ^2	DF	χ^2/DF	P
1	0.02	1.045	184.30	0.765	1.164	2	0.582	0.805	16.510	1	16.51	0.000
2	0.04	2.230	195.26	0.562	0.123	5	0.0246	1.000	7.559	4	1.890	0.109
3	0.1	5.175	192.25	0.621	4.143	8	0.518	0.844	0.787	7	0.112	0.998
4	0.8	41.700	190.41	0.657	5.825	7	0.832	0.560	5.390	6	0.898	0.495
5	1.0	54.205	208.37	0.310	10.312	9	1.145	0.326	15.367	8	1.921	0.052
6	2.0	106.36	184.22	0.766	5.148	13	0.396	0.972	5.009	12	0.417	0.958

C

Run	Dwell Time (s)	Mean	Standard Deviation	Measurements in $<n>\pm x^*\sigma$, (% of total) x=1	x=1.5	x=2	x=3
1	0.02	1.045	0.981	114 (57.0%)	182 (91.0%)	197 (98.5%)	197 (98.5%)
2	0.04	2.230	1.475	140 (70.0%)	163 (81.5%)	195 (97.5%)	199 (99.5%)
3	0.1	5.175	2.230	147 (73.5%)	176 (88.0%)	191 (95.5%)	200 (100.0%)
4	0.8	41.700	6.301	135 (67.5%)	172 (86.0%)	195 (97.5%)	200 (100.0%)
5	1.0	54.205	7.515	137 (68.5%)	175 (87.5%)	192 (96.0%)	200 (100.0%)
6	2.0	106.360	9.898	134 (67.0%)	174 (87.0%)	194 (97.0%)	200 (100.0%)

Figure 2.6 Statistical Analyses and Tests on Data. A. Tabular results of analyses showing sample mean for each run, variance, standard deviation and standard deviation of the mean or standard error and the Poisson estimates for standard deviation and standard error. B. χ^2 and χ^2 probability of the mean P calculated for samples and for Poisson and Gaussian distributions. C. Comparison of the experimental distribution with the expected Gaussian or normal distribution, showing number of measurements occurring within the standard error limits of $\pm 1\sigma$, $\pm 1.5\sigma$, $\pm 2\sigma$, $\pm 3\sigma$. Data was taken using a NaI(Tl) scintillation detector with multi-channel analyzer set for multi-scaling, by Creighton University NIM students, Fall 2019, analyzed and tabulated by Zachary J. Sabata (Lab Partners: Jeffrey Wong, Alex Marvin and Caleb Thiegs)

To illustrate these steps clearly, we show in Fig. 2.6 the results of such steps performed on real data taken using a NaI(Tl) scintillation detector with multi-channel analyzer set for multi-scaling, by Creighton University NIM students, Fall 2019. Counts from a Co-57 source were recorded for selected dwell times, here, 0.02 s, 0.04 s, 0.1 s, 0.8 s, 1 s and 2 s. The dwell times were chosen in view of the activity of the radioactive source, such that the average counts ranged from hundreds to 1 or

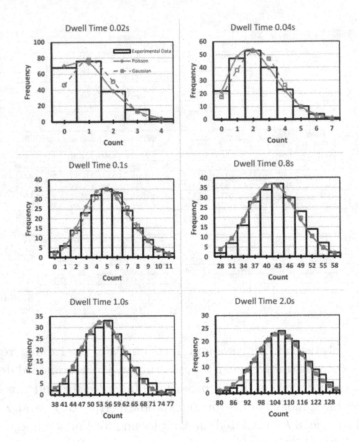

Figure 2.7 Comparisons of experimental frequency distribution histograms with Poisson (orange or balls) and Gaussian (blue or squares) distribution function predictions. Data was taken using a NaI(Tl) scintillation detector with multi-channel analyzer set for multi-scaling, by Creighton University NIM students, Fall 2019, analyzed and plotted by Zachary J. Sabata (Lab Partners: Jeffrey Wong, Alex Marvin and Caleb Thiegs).

2. Sample experimental results for runs with dwell times 0.02 s and 0.04 s are shown in Fig. B.1 in the Appendix, where raw counts were binned into cells with specified count range and their frequencies. The Poisson and Gaussian predicted values were calculated using Eq. 2.33 and Eq. 2.44 respectively. The Chi-Squared values were obtained using Eq. 2.82 based on the residuals shown in the table therein. Independent cell numbers were assigned only for cells containing at least 5 measurements and

those below 5 were assigned the next lower cell number having 5 or more measurements, in accord with the assumptions of the distributions being tested [51].

Fig. 2.6A shows in tabular form, results of analyses based on the constructed distribution tables (see samples in Fig. 2.7). These include sample mean $\bar{x}$ for each of the six runs calculated using Eq. 2.79, or Eq. 2.68; sample variance calculated using Eq. 2.71 and sample standard deviation obtained therefrom. The standard deviation of the mean or standard error was obtained as the square root of the variance of the sample mean, Eq. 2.75. The Poisson estimates for standard deviation σ_P and standard deviation of the mean or standard error $\sigma_{<n>}$ were calculated using the assumptions in Eq. 2.36 and Eq. 2.38, namely, that the variance is the mean, with the sample mean taken as the best estimate of the mean. The Poisson estimates of standard deviation are very close to values obtained as the square root of the variance of the sample mean.

Fig. 2.6B, shows the χ^2 and χ^2 probability P calculated for the samples and for Poisson and Gaussian distributions. For each run, χ^2 of the data (mean) was calculated using Eq. 2.51, while χ^2 for Poisson and Gaussian were done using Eq. 2.82 with $DF = N - 2$ for the Poisson distribution and $DF = N - 3$ for the Gaussian distribution. Note that N here refers to the number of cells for the χ^2 probability P for each of these three.

Fig. 2.6C shows the comparison of the experimental distribution with the expected Gaussian or normal distribution, displaying number of measurements occurring within the standard error limits of $\pm 1\sigma$, $\pm 1.5\sigma$, $\pm 2\sigma$, $\pm 3\sigma$.

2.10 UNCERTAINTIES: CALCULATION AND EXPRESSION

2.10.1 Accuracy and Precision, Error and Uncertainty

For a statement of the result of a physical measurement to be meaningful, it must have two essential components, namely, a numerical value of the quantity measured in a specified system of units and the uncertainty associated with this numerical value. The uncertainty is an indication of the **accuracy** and **precision** of the result. Accuracy is the closeness of agreement between a measured value and the true value. The level of accuracy is indicated by stating the **measurement error**. As illustrated in Fig. 2.8, measurement error is the difference between a measurement and the true value of the quantity being measured (sometimes called

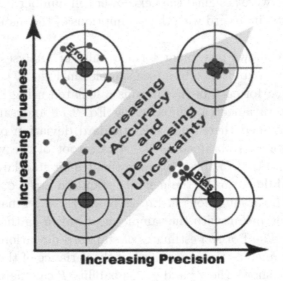

Figure 2.8 Graphical Illustration of Precision versus Accuracy, Error versus Uncertainty. Image adapted from NSF Advanced Technological Education Advanced Technological Education.

the **measurand**). Thus, measurement errors have numbers ascribed to them. Although mistakes or measurement blunders are also measurement errors, they are not used in stating uncertainty since they warrant data rejection. Measurement blunders are sometimes called **personal errors**. The measurement errors used in stating uncertainty are *systematic errors* and *random or statistical errors*. Systematic errors relate to the accuracy of a measurement while random errors are related to the precision of a measurement. The lower the systematic errors, the higher the accuracy (see Fig. 2.8). Likewise, the lower the random errors, the higher the precision (see Fig. 2.8 and Fig. 2.2). The total error is usually a combination of systematic error and random error. Some times, the results are quoted with two errors. The first error quoted is usually the random or statistical error, and the second is the systematic error. For example a measurement of distance might yield:

$$x = 1.93 \pm 0.06(stat.) \pm 0.04(syst.) \text{ m}.$$

If only one error is quoted it is the combined error.

The **bias** is the difference between the expected value of the results from estimates or measurements and the underlying true value.

Finally, uncertainty is the parameter that characterizes the range within which the values from measurements lie within a specified confidence level. Systematic and random uncertainties exist and should be estimated separately. When stated, the uncertainty gives the accuracy level of the results.

2.10.2 Statistical and Systematic Uncertainties

2.10.2.1 Statistical or Random Uncertainties

Random fluctuations or statistical errors lead to the following properties of statistical uncertainties [115].

1. They can be reliably estimated by repeating measurements

2. They follow a known distribution such as a Poisson distribution or a normal distribution or they can be determined empirically from the distribution of an unbiased, sufficiently large sample.

3. Their relative or fractional uncertainty reduces as $1/\sqrt{N}$ where N is the sample size

4. Examples include radioactive decay which follows a Poisson distribution; the efficiency of a detector which follows a binomial distribution

2.10.2.2 Systematic Uncertainties

1. Cannot be calculated solely from sampling fluctuations

2. In many cases they do not reduce as $1/\sqrt{N}$ (but often also become smaller with larger N)

3. They are difficult to determine and are generally less well known than the statistical uncertainty

4. Examples include background noise, calibration uncertainties, uncertainties due to detector resolution

2.10.3 Calculation, Estimation and Expression of Uncertainties

The uncertainties give the range of the errors, thereby indicating the probability of finding the true mean μ. For instance, if a single measurement of x yields a count of $n = 100$ and the process is assumed to be a random process whose errors are normally distributed, the best estimate of the standard deviation for this single measurement is given by:

$$\sigma = \sqrt{n} = \sqrt{100} = 10,$$

in accord with Eq. 2.46. If the statistical model chosen is the exponential distribution, then the best estimate of σ for x is given by:

$$\sigma = n = 100,$$

in accord with Eq. 2.60. However, if a Poisson process is involved, then the best estimate of σ for x is

$$\sigma = \sqrt{n} = 10,$$

in accord with Eq. 2.36. A statistical uncertainty σ represents some kind of probability distribution which is very often a Gaussian. It is usual to assume the uncertainty function as a Gaussian if nothing else is stated. Furthermore, unless stated otherwise, the interval $x \pm \sigma$ corresponds to 1σ errors or a probability of 68%. In general, a single measurement x has its error range or uncertainty stated as $x \pm \sigma$ and the σ is given by the assumed probability distribution for the measurement. Furthermore, using the normally distributed case for $n = 100$ for x, the error intervals are given in Table 2.3.

Thus, the conventional expression of the result of measuring x is to state it as $x \pm 1\sigma$ and this interval is expected to contain the true mean μ with a probability of 68%. To increase this probability to 99.7%, one can state x as $x \pm 3\sigma$.

2.11 ERROR PROPAGATION

2.11.1 Error Propagation Formula

If we measure several random variables such as radioactive counts X_1, X_2 and ...,X_3 we may find from our experimental data quantities such as $\bar{X}_1$, $\bar{X}_2$ and $\bar{X}_3$; and can then find their corresponding variances σ_1^2, σ_2^2 and σ_3^2. However, in many cases, further data analysis provides formulas

Table 2.3 Error Intervals and Probability for a Single Measurement of x 100 counts, assuming Normal or Gaussian Error Distribution. Image adapted from Nuclear Medicine Physics: A Handbook for Teachers, p171. IAEA 2014.

Interval relative to σ	Interval (values)	Probability that true mean is included (%)
$x \pm 0.67\sigma$	$96.3 - 106.7$	50
$x \pm 1\sigma$	$90.0 - 110.0$	68
$x \pm 1.64\sigma$	$83.6 - 116.4$	90
$x \pm 2\sigma$	$80.0 - 120.0$	95
$x \pm 2.58\sigma$	$74.2 - 125.8$	99
$x \pm 3\sigma$	$70.0 - 130.0$	99.7

involving the X_1, X_2 and ..., X_m so that we encounter functions which we write as

$$y = f(X_1, X_2, ..., X_m). \tag{2.84}$$

Very often we need to find the variance σ_y^2 of this arbitrary function $y(X)$. We start with a Taylor expansion of $y(X)$ around the mean values, $\bar{X}_1$, $\bar{X}_2$ up to $\bar{X}_m$:

$$y = f(\bar{X}_1, \bar{X}_2, ..., \bar{X}_m) + \left(\frac{\partial f}{\partial X_1}\right)_{\bar{X}} (X_1 - \bar{X}_1)$$

$$+ ... \left(\frac{\partial f}{\partial X_m}\right)_{\bar{X}} (X_m - \bar{X}_m) + ... \tag{2.85}$$

We have truncated the expansion with first order terms since the y is assumed to be a linear function of the X_1, X_2, ..., X_m or at least a linear function of the $\bar{X}_1$, $\bar{X}_2$, ..., $\bar{X}_m$. Note that

$$\left(\frac{\partial f}{\partial X_m}\right)_{\bar{X}}$$

is just the evaluation of the partial derivative at the point of expansion: $\bar{X}_1$, $\bar{X}_2$, ..., $\bar{X}_m$. Using the explicit definition of σ_y^2, we have:

$$\sigma_y^2 = \frac{1}{n} \sum_{i=1}^{n} (y - \bar{y})^2) \tag{2.86}$$

$$\sigma_y^2 = \frac{1}{n} \sum_{i=1}^{n} \left[\left(\frac{\partial f}{\partial X_1}\right)_{\bar{X}} (X_1 - \bar{X}_1) + ... \left(\frac{\partial f}{\partial X_m}\right)_{\bar{X}} (X_m - \bar{X}_m) + ... \right]^2. \tag{2.87}$$

Of course, Eq. 2.85 implies that

$$\bar{y} = f\left(\bar{X}_1, \bar{X}_2, ..., \bar{X}_m\right)$$

and so the averages, that is, the first terms in Eq. 2.87 contribute zero to the variance of y, σ_y^2. Expanding the terms in the square brackets and continuing the summation, Eq. 2.87 produces the general expression for variance in an arbitrary function y as

$$\sigma_y^2 = \left(\frac{\partial f}{\partial X_1}\right)_{\bar{X}}^2 \frac{1}{n}\sum_{i=1}^{n}(X_1 - \bar{X}_1)^2 + \left(\frac{\partial f}{\partial X_2}\right)_{\bar{X}}^2 \frac{1}{n}\sum_{i=1}^{n}(X_2 - \bar{X}_2)^2 \, ...$$

$$\left(\frac{\partial f}{\partial X_m}\right)_{\bar{X}}^2 \frac{1}{n}\sum_{i=1}^{n}(X_m - \bar{X}_m)^2 \qquad (2.88)$$

$$+ 2\left(\frac{\partial f}{\partial X_1}\right)_{\bar{X}}\left(\frac{\partial f}{\partial X_2}\right)_{\bar{X}} \frac{1}{n}\sum_{i=1}^{n}(X_1 - \bar{X}_1)(X_2 - \bar{X}_2) + ...$$

More compactly, we have:

$$\sigma_y^2 = \left(\frac{\partial f}{\partial X_1}\right)_{\bar{X}}^2 \sigma_1^2 + \left(\frac{\partial f}{\partial X_2}\right)_{\bar{X}}^2 \sigma_2^2 + ... \left(\frac{\partial f}{\partial X_m}\right)_{\bar{X}}^2 \sigma_m^2$$

$$+ 2\left(\frac{\partial f}{\partial X_1}\right)_{\bar{X}}\left(\frac{\partial f}{\partial X_2}\right)_{\bar{X}} \sigma_1 \sigma_2 + ..., \qquad (2.89)$$

where σ_m^2 is the variance of X_m and $\sigma_1 \sigma_2$ is the co-variance of X_1 and X_2 as defined in Eq. 2.18. When the variables X_m are mutually unrelated, as is often the case, then their errors and variances are uncorrelated, rendering the co-variance terms in Eq. 2.89 as zero. Hence, for mutually unrelated random variables X_m, the variance of their function $y = f(X_m)$ is given by:

$$\sigma_y^2 = \left(\frac{\partial f}{\partial X_1}\right)_{\bar{X}}^2 \sigma_1^2 + \left(\frac{\partial f}{\partial X_2}\right)_{\bar{X}}^2 \sigma_2^2 + ... \left(\frac{\partial f}{\partial X_m}\right)_{\bar{X}}^2 \sigma_m^2, \qquad (2.90)$$

or even more compactly:

$$\sigma_y^2 = \sum_{i=1}^{m}\left(\frac{\partial f}{\partial X_i}\right)_{\bar{X}}^2 \sigma_i^2. \qquad (2.91)$$

Using Eq. 2.90, we show in Table 2.4 the variances for various common functions, with $f(X_1, X_2...)$ now rendered as $f(x, y)$ for convenience:

Table 2.4 Some Error Propagation Formulas

Function	Partial Derivatives	Variance
$f = kx$	$\frac{\partial f}{\partial x} = k$	$\sigma_f^2 = k^2 \sigma_x^2$
$f = x+y$	$\frac{\partial f}{\partial x} = 1;\ \frac{\partial f}{\partial y} = 1$	$\sigma_f^2 = \sigma_x^2 + \sigma_y^2$
$f = x-y$	$\frac{\partial f}{\partial x} = 1;\ \frac{\partial f}{\partial y} = -1$	$\sigma_f^2 = \sigma_x^2 + \sigma_y^2$
$f = xy$	$\frac{\partial f}{\partial x} = y;\ \frac{\partial f}{\partial y} = x$	$\left(\frac{\sigma_f}{f}\right)^2 = \left(\frac{\sigma_x}{x}\right)^2 + \left(\frac{\sigma_y}{y}\right)^2$
$f = \frac{x}{y}$	$\frac{\partial f}{\partial x} = \frac{1}{y};\ \frac{\partial f}{\partial y} = \frac{-x}{y^2}$	$\left(\frac{\sigma_f}{f}\right)^2 = \left(\frac{\sigma_x}{x}\right)^2 + \left(\frac{\sigma_y}{y}\right)^2$

2.11.2 Examples of Error Propagation

Example 2.16 *In a typical alpha spectroscopy experiment aimed at determining the absolute activity of an alpha-particle emitting nuclide, the following formula can be used:*

$$A = \frac{\Sigma}{t} \frac{4\pi d^2}{a}, \tag{2.92}$$

where A is the activity, Σ is the number of counts recorded by the detector, d is the distance from the source to the detector and t is the duration of measurement by the detector. Derive the expression for the standard deviation σ_A for the activity.

Solution

Using Eq. 2.84, the activity A is a function y such that

$$y = f(X_1, X_2, ..., X_m) = A(\Sigma, t, d, a) \tag{2.93}$$

Thus, the variance in A, σ_A^2 can be derived from Eq. 2.90 as:

$$\sigma_A^2 = \left(\frac{\partial A}{\partial \Sigma}\right)_{\bar{\Sigma}}^2 \sigma_\Sigma^2 + \left(\frac{\partial A}{\partial t}\right)_{\bar{t}}^2 \sigma_t^2 + \left(\frac{\partial A}{\partial a}\right)_{\bar{a}}^2 \sigma_a^2 + \left(\frac{\partial A}{\partial d}\right)_{\bar{d}}^2 \sigma_d^2. \tag{2.94}$$

And finally, the standard deviation in activity σ_A is obtained as:

$$\sigma_A = A \sqrt{\left(\frac{\sigma_\Sigma}{\Sigma}\right)^2 + \left(\frac{\sigma_t}{t}\right)^2 + \left(\frac{\sigma_a}{a}\right)^2 + \left(2\frac{\sigma_d}{d}\right)^2}. \tag{2.95}$$

Example 2.17 *In an alpha particle spectroscopy experiment carried out by students, Polonium 210 as source and a surface barrier detector with a theoretical detection efficiency of 100% were used. If 7789 counts were detected in 6223 seconds, find (a) the absolute activity A of the Po-210 sample and (b) standard deviation σ_A of the measured activity. The source to detector distance d was measured and found to be $53 \pm 1\,mm$. The surface area of the detector was $25 \pm 1\,mm^2$.*

Solution

(a) Using Eq. 2.92, we have

$$A = \frac{\Sigma}{t} \frac{4\pi d^2}{a} = \frac{7789}{6223} \frac{4\pi (0.053)^2}{2.5 \times 10^{-5}} = 1.8 \times 10^3 \, \text{dps}, \qquad (2.96)$$

where dps is disintegrations per second.
The activity is thus $0.04865\,\mu Ci$.

(b) Note that the uncertainty in time can be estimated to be about 1 s. The uncertainty in the counts is the square root of the counts, following Poisson statistics or the random nature of radioactivity. The standard deviation in activity σ_A is calculated using Eq. 2.95 as follows:

$$\sigma_A = A \sqrt{ \left(\frac{\sigma_\Sigma}{\Sigma} \right)^2 + \left(\frac{\sigma_t}{t} \right)^2 + \left(\frac{\sigma_a}{a} \right)^2 + \left(2 \frac{\sigma_d}{d} \right)^2 }$$

$$= 1800 \sqrt{ \left(\frac{88}{7789} \right)^2 + \left(\frac{1}{6223} \right)^2 + \left(\frac{1}{25} \right)^2 + \left(2\frac{1}{53} \right)^2 }$$

$$= 101 \, dps \qquad\qquad\qquad\qquad\qquad\qquad = 0.00273 \, \mu Ci.$$

The activity is therefore $1800 \pm 101\,dps$ or $0.049 \pm 0.003\,\mu Ci$.

Example 2.18 *A student is attempting to experimentally check the efficiency ε of a new detector in order to confirm the manufacturer's stated value. She uses a radioactive counting standard with a decay rate of 3000 ± 50 decays per minute and measures a decay rate of 900 ± 40 decays per minute. (a) What is the measured efficiency of the counting system? (b) What is standard deviation of the measurement?*

Solution

(a) The efficiency of the detector is given by Eq. 2.97:

$$\varepsilon = \frac{\text{detected counts}}{\text{total decays from source}}. \qquad (2.97)$$

Thus,

$$\varepsilon = \frac{900}{3000} = 0.3 = 30\%.$$

(b) σ_ε is calculated using Eq. 2.95 as follows:

$$\sigma_\varepsilon = \varepsilon \sqrt{\left(\frac{\sigma_{Det}}{Det}\right)^2 + \left(\frac{\sigma_{Source}}{Source}\right)^2}$$

$$= 0.3 \sqrt{\left(\frac{50}{3000}\right)^2 + \left(\frac{40}{900}\right)^2}$$

$$= 0.009$$

Exercise 26

A student in a NIM lab measures the activity of a radioactive sample and obtains a mean count of 10,000 for a specified duration of counting. (a) What is the standard deviation? (b) From the expected distribution, 68% of the counts will fall between what range?

Exercise 27

Two radioactive sources were monitored with a detector at equal collection times. Assuming source A produced a net count of 18,000 and source B produced a net count of 12,000, what is the ratio of Activities (A/B) of the sources, including the appropriate uncertainty. Hint: consult Table 2.4 for the error propagation with $f = x/y$.

Exercise 28

In 10 minutes, a Geiger counter recorded 500 counts during an environmental survey. Measurement of background for 60 minutes yielded 1440 counts. Find the net counting rate and its standard deviation, using appropriate error propagation formulas.

Exercise 29

During radiation measurements, background counts N_b which do not come from the sample or target volume but which may arise from electronic noise, detection of cosmic rays, natural radioactivity in the detector and even scatter radioactivity from non-target radionuclides in the sample, may contribute to the total or gross counts N_g such that the net sample count N_s becomes:

$$N_s = N_g - N_b.$$

(a) Use appropriate error propagation formulas (from Table 2.4) to show that the standard deviation for the net sample count σ_{N_s} is given by

$$\sigma_{N_s} = \sqrt{N_g + N_b}. \tag{2.98}$$

(b) Use Eq. 2.39 to show that the fractional standard deviation or coefficient of variation CV is given by:

$$CV_{N_s} = \frac{\sqrt{N_g + N_b}}{N_g - N_b}. \tag{2.99}$$

(c) For $N_g = 570$ and $N_b = 180$, find (i) N_s, (ii) σ_{N_s} and (iii) CV_{N_s}.

Exercise 30

In radiation measurements, the background count rate r_b and gross rate r_g lead to the following expression for the net sample count rate r_s:

$$r_s = r_g - r_b.$$

(a) Assuming time is kept so accurately that the uncertainty is negligible during radiation measurements such that it can be treated as a constant or scaling factor, use Eq. 2.42 and appropriate error propagation formulas (from Table 2.4) to show that the standard deviation for the net sample count rate σ_{r_s} is given by

$$\sigma_{r_s} = \sqrt{\frac{r_g}{t_g} + \frac{r_b}{t_b}}. \tag{2.100}$$

(b) Use Eq. 2.39 to show that the fractional standard deviation or coefficient of variation CV for the sample count rate is given by:

$$CV_{r_s} = \frac{\sqrt{\frac{r_g}{t_g} + \frac{r_b}{t_b}}}{r_g - r_b}. \tag{2.101}$$

For $r_g = 2850$ and $r_b = 900$ both measured for 5 minutes, find (i) r_s, (ii) σ_{r_s} and (iii) CV_{r_s}.

Exercise 31

A radioactive source gives 1600 gross counts in 1 minute. The background is counted for 10 minutes and produces 121 counts. Find (a) the net sample count rate and (b) the net sample standard deviation.

Exercise 32

In many nuclear experiments involving radioactivity, the difference between two measurements is considered significant if that difference is greater than 3σ. Hence, the minimum net sample counts N_s that can be detected with 0.3% confidence is given by $N_{smin} = N_1 - N_2 = 3\sigma_{N_{smin}} = 3\sqrt{N_1 + N_2}$. Let N_1 be the gross counts N_g and N_2 be the background counts N_b. (a) Show by minimization that the minimum N_g required to give the minimum detectable counts N_{smin} with 0.3% confidence is given by

$$N_g = \frac{(2N_b + 9) + \sqrt{(72N_b + 81)}}{2} \tag{2.102}$$

Note that after getting above N_g, then one can get $N_{smin} = N_g - N_b = 3\sigma_{N_{smin}} = 3\sqrt{N_g + N_b}$. From this N_{smin}, minimum detectable activity (MDA) can be calculated. (b) A GM counter is used to obtain N_b as 540 counts in 5 minutes. Find (i) N_g needed to obtain a minimum detectable net counts, N_{smin}. (ii) What is N_{smin}? Hint: use Eq. 2.102.

Exercise 33

To obtain the optimal times for counting background t_b and gross t_g in order to minimize the statistical uncertainty of the net sample counting rate, Eq. 2.100 is squared to get the variance and then minimized. Show that minimization leads to the following expression for the minimum total counting time t, where $t = t_g + t_b$:

$$\frac{t_g}{t - t_g} = \sqrt{\frac{r_g}{r_g}} \tag{2.103}$$

Exercise 34

In a single measurement of a radioactive source, you detect 10,000 counts and quickly note $\sigma = \sqrt{10,000} = 100 \ counts$. What is the interval of counts for: (a) a 68% probability that the true mean is within that interval? (b) a 95% probability that the true mean is within that interval? Hint: you may use Table 2.3.

Exercise 35

A radioactive source is found to have a count rate of 4 counts per second. What is the probability of observing (a) zero counts in a period of 2 seconds; (b) 2 counts in a period of 4 seconds? Hint: use Poisson distribution with $\mu = r(t) \times t$ and $n = counts$, Eq. 2.33, to find P(n).

Exercise 36

You set up a detector to measure a radioactive sample that is completely shielded from background radiation. How many counts should you record in order to have a CV or fractional standard error of 1% with a confidence interval of 95%? Hint: 95% interval requires 2σ. Assume Poisson statistics with $\sigma = \sqrt{N}$ and solve for N.

Exercise 37

You are interested in using just two measurements of a radioactive sample to check if a radiation detection equipment is working as expected. Since the standard error of the difference between means from two samples, σ_{diff} based on error propagation (see Table 2.4) can be written thus:

$$\sigma_{diff} = \sqrt{\sigma_{\mu_1}^2 + \sigma_{\mu_2}^2}, \tag{2.104}$$

where μ_1 and μ_2 refer to the two means. Applying Eq. 2.100 to Eq. 2.104, you arrive at the equation behind the well-known t-test or tail-test which is the number of standard deviations from zero:

$$t = \frac{|\mu_1 - \mu_2|}{\sqrt{\frac{r_1}{t_1} + \frac{r_2}{t_2}}}. \tag{2.105}$$

Now, your μ_1 was 45,123 cpm, in a 5-minute measurement and μ_2 was 44,916 cpm, done in a 2-minute check. Find t using Eq. 2.105 and state if the detector was functioning as expected.Hint $t = 2$ implies 2 σ or that 5% of the time (95% confidence interval), such a difference comes from the same sample and $t = 1$ implies 1 σ or that 32% of the time (68% confidence interval), such a difference comes from the same sample.

2.12   ANSWERS

Answer of exercise 23

You have been guided to the proof already in the exercise.

Answer of exercise 24

Steps have already been given in the text.

Answer of exercise 25

Ans (a) Using Eq. 2.47, $\sigma = \frac{FWHM}{2.35} = 31.68$ keV.

Answer of exercise 26

Ans (a) S D or $\sigma = \sqrt{mean} = 100$. (b) 68% falls between $+\sigma$ $-\sigma =$ 10100 and 9,900.

Answer of exercise 27

Solution 1.5 ± 0.0176.

Answer of exercise 28

Rate $= 500/10 - 1440/60 = 50\text{--}24$ cpm $= 26$ cpm $\sigma = \sqrt{(50/5 + 24/60)} = \sqrt{20.4} = 4.52$. Hence, net counting rate $= 26 +/- 4.5$ cpm.

Answer of exercise 29

Ans (c) (i) 390 counts. (ii) 27.4 counts. (iii) 0.07 or 7%

Answer of exercise 30

Ans (c) (i) 1950 counts per minute. (ii) 27.4 counts per minute. (iii) 0.014 or 1.4%

Answer of exercise 31

Ans (a) $R_n = R_g - R_b = 1587.9$ cpm. (b) $\sigma = 40.15$.

Answer of exercise 32

Ans: (i) $N_g = 643$. (ii) $N_{smin} = 103$ counts.

Answer of exercise 33

Steps already given in text.

Answer of exercise 34

Ans (a) 9900-10100 (mean $+/-1$ σ) (b) 9800-10200 (mean $+/- 2\sigma$)

Answer of exercise 35

Ans (a) $P(0) = ((8^0)e^{(-8)})/0! = 3.35 \times 10^{-4}$ or $3.35E^{-4}$ (b) $P(2) = ((8^2)e^{(-8)})/2! = 0.01$.

Answer of exercise 36

Ans $2\sigma/N = 1\%$ implies $N = (2/1\%)^2 = 40,000$ or greater.

Answer of exercise 37

Ans $\mu_1 - \mu_2 = 207$. So $t = \frac{207}{177.43} = 1.17$. Thus, 1.17σ. At 68% confidence interval, this is acceptable. However, often, a 95% confidence interval is preferred for equipment t-test.

CHAPTER 3

Energy Loss of Heavy Charged Particles through Matter

Here, we consider electrons and positrons to be light charged particles. Heavy charged particles are those with rest masses much greater than electrons. They include protons, muons, pions, α particles and other light nuclei and heavy ions[1]. This chapter considers the energy loss mechanisms of heavy charged particles as they interact with matter (see Fig. 3.1). Generally, as heavy charged particles pass through materials, they experience two things: energy loss and change in direction (scattering). These two experiences are the result of at least six interactive processes [2] between the charged particles and the materials. We focus on the Mathematics used in keeping track of these two main effects: energy loss and scattering for heavy charged particles.

3.1 GENERAL RESULTS AND PERTURBATION THEORY

Let us give the general result and then present some of the mathematics behind the derivations. Energy loss and change in direction of charged

---

[1]We will explicitly include heavy ions from the mathematical treatment here although the additional effects due to them often engender their being excluded from simple treatments of heavy charged particles.

[2]1. Inelastic collisions with atomic electrons of the material; 2. Elastic scattering from the nuclei of the material; 3. Bremsstrahlung; 4. Nuclear reactions; 5. Emission of Cherenkov radiation and 6. Transition radiation.

DOI: 10.1201/9781003215622-3

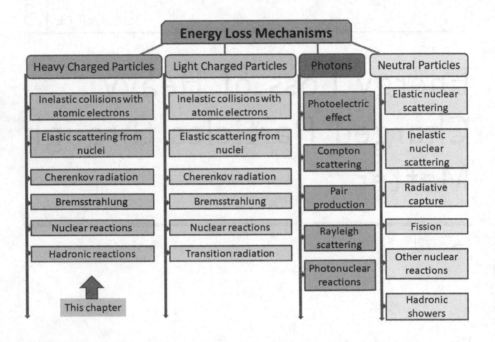

Figure 3.1 Energy Loss Mechanisms for Heavy Charged Particles, Light Charged Particles, Photons and Neutral Projectiles. Properties of the projectiles such as charge, mass, velocity and energy determine the relative predominance of the processes. This chapter deals with heavy charged particles.

particles as they traverse matter are largely due to their inelastic collisions with atomic electrons of the material. Such inelastic collisions are electromagnetic interactions which in general are currently described by the Bethe stopping power formula due to Hans Bethe [11, 13][3]. The *stopping power S* of a material is defined as the mean energy loss per distance travelled by charged particles (eg, protons, alpha particles, atomic ions) traversing that material. So we speak of the *stopping power* of water for alpha particles or the *mean energy loss* of alpha particles in water. For a charged particle with charge number Z_1 travelling with energy E, through a random target material with atomic number Z_2 and average

_____

[3]Also called Bethe-Bloch Formula, somewhat inaccurately, since Bloch gave comparisons with Bohr's formula and showed conditions of agreement with Bohr's [128]

electron density $n_e = NZ_2$, the energy loss formula has a general form in SI units [J/m]:

$$-\frac{dE}{dx} = \frac{Z_1^2 e^4 n_e}{4\pi\varepsilon_0^2 m_e v^2}L = \frac{Z_1^2 e^4 NZ_2}{4\pi\varepsilon_0^2 m_e v^2}L,$$ (3.1)

where m_e is the mass of the electron. The negative sign in Eq. 3.1 indicates that it gives the energy *lost* by the particle as it traverses matter. The factor L is the so-called *stopping logarithm* (or *stopping number* [128], [69] which encapsulates classically and quantum mechanically-derived expressions of the energy loss. The most important experimental takeaway from Eq. 3.1 is that the rate of energy loss of charged projectiles in a material due to excitation and ionization inside the material increases rapidly with decreasing velocity (v^2) and with increasing charge of the projectile (Z_1^2). S also depends on the density of the material because it is proportional to the number of atoms per volume. Thus, S is directly proportional to the absorber's density ρ and inversely proportional to its atomic mass A_2. In spite of increasing use of SI units, it is still common to do calculations of stopping power S in units of A MeV cm^{-1}. Note that "A MeV" is read "MeV per nucleon". Other units include eV nm^{-1}. Also, in practice, the stopping power is usually divided by the density of the random target material to obtain the *mass stopping power*, S/ρ given in units of A MeV g^{-1}cm^2 by:

$$\frac{S}{\rho} = \frac{1}{\rho}\frac{dE}{dx} = 4\pi m_e c^2 r_e^2 \frac{Z_1^2 N_A Z_2}{A_2 \beta^2}L = (4\pi m_e c^2 r_e^2)\frac{Z_1^2}{\beta^2}n_e L.$$ (3.2)

Here, A_2 is the atomic mass (or effective atomic mass) of the absorber, n_e is the electron density of the absorber (number of electrons/g) and r_e, the classical electron radius which has the usual form:

$$r_e = \frac{e^2}{4\pi\varepsilon_0 m_e c^2} = 2.818 \times 10^{-15}\,\text{m} = 2.818\,\text{fm}.$$ (3.3)

The quantity in parenthesis in Eq. 3.2 is a constant and therefore independent of the charged particle properties. It is also independent of the absorber or stopping medium. For convenience in calculations, it is expressed as the Mass Collision Stopping Constant, C_1, given by:

$$C_1 = 4\pi m_e c^2 r_e^2 = 5.099 \times 10^{-25}\,\text{MeV.cm}^2.$$ (3.4)

Other units of the mass stopping power include eV μg^{-1}cm^2 and eV $atom^{-1}cm^2$. In experimental work, *specific ionization* is also used to describe energy loss of charged particles. The specific ionization is defined as the number of ion pairs (i.p) formed per unit distance travelled by the charged projectiles:

$$\text{Specific Ionization} = \frac{dE/dx\,[\text{eV/cm}]}{w\,[\text{eV}/i.p]}, \tag{3.5}$$

where w is the the average energy needed to create an ion or electron-hole pair. Of course, the energy required to create an ion pair is material-dependent. A graph of specific ionization as a function of residual energy can be used to display the so-called Bragg peak for charged projectiles in a material and such a graph is dimensionally equivalent to a graph of energy loss as a function of the kinetic energy of the projectile which we show later while treating Bragg curve (Fig. 3.4).

Example 3.1 *Calculate the electron density of water n_e in units of electrons/g, in view of the occurrence of n_e in Eq. 3.1. Hint: 1 mole of water H_2O is 18.0153 g of water.*

Solution

From the 2 hydrogen atoms and 1 oxygen atom that make one molecule of water, $Z_2 = 10$ electrons/molecule. The mass of 1 mole of water is 18.0153 g and contains N_A entities (which could be atoms, electrons, molecules, particles). Therefore, for 1 g of water, we have,

$$n_e = (\frac{1\,\text{mole}}{18.0153\,\text{g}})(6.022 \times 10^{23}\,\frac{\text{molecules}}{\text{mole}})(10\,\frac{\text{electrons}}{\text{molecule}})$$

$$= 3.343 \times 10^{23}\,\frac{\text{electrons}}{g}.$$

Note that $n_e = NZ_2$ in electrons/g is also given by:

$$n_e = \frac{N_A}{A_2}Z_2, \tag{3.6}$$

where N_A is Avogadro's number, A_2 is the atomic mass of the target material (or effective atomic mass in the case of molecules) and this expression (Eq. 3.6) was used to simplify Eq. 3.2. Succinctly, the mass stopping power $\frac{S}{\rho}$ is proportional to the electron number density of the material which depends on the ratio $\frac{Z_2}{A_2}$ and this ratio is about 0.5 for most materials apart from Hydrogen for which $\frac{Z_2}{A_2} = 1$. More precisely, for all other elements, $\frac{Z_2}{A_2}$ slowly decreases from 0.5 for low Z elements to about 0.38 for high Z elements.

Different forms of L in Eq. 3.1 and Eq. 3.2 enable classically or quantum mechanically derived expressions for the energy loss or stopping power. Mathematically, perturbation theory has been the main method for the quantum mechanical derivations. Bethe used first order perturbation theory (first order Born approximation [20]) starting from the time-dependent Schrödinger equation [106] for the interaction:

$$i\hbar \frac{d\Psi(\mathbf{r},t)}{dt} = (H + V_{int}(\mathbf{r},t))\Psi(\mathbf{r},t). \tag{3.7a}$$

$$V_{int}(\mathbf{r},t) = \sum_{i=1}^{Z_2} \frac{-e_1 e}{(\mathbf{r}_i - \mathbf{R}(t))}. \tag{3.7b}$$

$$\Psi(\mathbf{r},t) = \sum_j c_j(t) e^{-i\varepsilon_j t} |j\rangle. \tag{3.7c}$$

In Eq. 3.7a, H is the Hamiltonian of the unperturbed system, precisely, an isolated atom of the target material. That is, $H = H_0 + V_0$ while V_{int} is the perturbation or interaction potential, defined in Eq. 3.7b, where r_i is the position operator of the i_{th} electron. $\mathbf{R}(t)$ gives the projectile's trajectory. The time dependent wave function for a bound state of the atom is expressed in terms of stationary waves as shown in Eq. 3.7c where $|j\rangle$ are solutions of the eigen-equation:

$$H|j\rangle = \varepsilon_j |j\rangle \tag{3.8}$$

Perturbation theory based on Born approximation implies that the c_j coefficients in Eq. 3.7c can be expressed as powers of the interaction or perturbation potential V_{int} thus:

$$c_j(t) = \delta_{j0} + c_j^{(1)}(t) + c_j^{(2)}(t) + \dots \tag{3.9}$$

Here,

$$\delta_{j0} = \begin{cases} 1 & \text{for } j = 0 \\ 0 & \text{for } j \neq 0 \end{cases}$$

and

$$c_j^{(1)}(t) = \frac{1}{i\hbar} \int_{-\infty}^{t} dt' e^{i\omega_{j0} t'} \langle j | V(\mathbf{r}, t') | 0 \rangle \tag{3.10}$$

$$c_j^{(2)}(t) = (\frac{1}{i\hbar})^2 \sum_k \int_{-\infty}^t dt' e^{i\omega_{jk}t'} \langle j|V(\mathbf{r},t')|k\rangle \, x$$

$$\times \int_{-\infty}^t dt'' e^{i\omega_{k0}t''} \langle k|V(\mathbf{r},t'')|0\rangle \, ... \qquad (3.11)$$

Only the terms $c_j^{(1)}(t = \infty)$ (Eq. 3.10) are important within the first order perturbation theory. Once the explicit expression for the potential is inserted in the $c_j^{(1)}(\infty)$s, Fourier transforms (eg, f($\mathbf{r}$) in coordinate space becomes f($\mathbf{q}$) in momentum space) and integrations are then performed to calculate the transition probabilities, P_j. We know from the postulates of quantum mechanics that

$$P_j = |\langle j|\Psi(\infty)\rangle|^2, \qquad (3.12)$$

which first order perturbation theory now provides as

$$P_j = |c_j^{(1)}(\infty)|^2. \qquad (3.13)$$

Bethe employed these steps in giving the first quantum-mechanically derived expression for the mean energy loss of charged particles and we will give the details after treating the classical formula that preceded Bethe's.

3.2 BOHR'S CLASSICAL FORMULA

3.2.1 Terminology and Physical Basis

Current work often terms the average rate of energy loss of a charged projectile per unit path length as it traverses a medium, **linear energy transfer**, LET. LET and energy loss pertain to the projectile. The energy received by the medium is the **stopping power** S and this can be less than LET because some of the LET energy may be removed from the track by fast secondary electrons, Cherenkov radiation and bremsstrahlung. The physical theories whose mathematical derivations we are delving into are actually theories of energy loss rate (thus, our chapter title). However, the literature is replete with the same theories being denoted as stopping power theories. The ICRU Report 90 [108] recommends restricted LET (or just LET), L_Δ and unrestricted LET L_∞ to demarcate the energy ranges (eg, L_{500} for energy losses less than 500 eV). Furthermore, connected with stopping power of a material S, often measured in MeV cm^{-1}, is the more practical quantity, mass stopping power, S/ρ often measured in MeV cm^2 g^{-1}. We note the current ICRU convention (ICRU 90 [108]) of connecting the linear stopping power S with the mean energy loss rate (in spite of the physical distinction we made above):

$$S = -\frac{dE}{dx} \tag{3.14}$$

where E is the mean energy lost by the charged projectile in traversing the distance dx in the target material. Again, the mass stopping power is then given by

$$\frac{S}{\rho} = \frac{1}{\rho}\frac{dE}{dx}. \tag{3.15}$$

In the derivations that follow, we need to keep in mind the physical phenomena being referred to. In general, the mean energy loss rate of a charged particle or stopping power of a material and therefore the mass stopping power of that material, has three physical components, namely, collisional or electronic stopping power S_{el}, radiative stopping power, S_{rad} and nuclear stopping power S_{nuc}. S_{el} is due to interactions of the charged projectile with atomic electrons of the target. S_{rad} is due to the emission of bremsstralung in the electric fields of the nuclei or electrons of the atoms of target materials. S_{nuc} arises from elastic Coulomb interactions wherein recoil energy is imparted to atoms of the target. Thus,

$$S = S_{el} + S_{rad} + S_{nuc}, \tag{3.16}$$

$$\frac{S}{\rho} = \frac{1}{\rho}S_{el} + \frac{1}{\rho}S_{rad} + \frac{1}{\rho}S_{nuc}. \tag{3.17}$$

The next section focuses on the mathematical derivation of average energy loss rate involved in S_{el}. The relative contributions of S_{el}, S_{rad} and S_{nuc} to S depends on the charge, mass and energy of charged projectile (as well as the properties of the medium) engendering our separation of the derivations for heavy charged particles and light charged particles. S_{el} dominates in the case of heavy charged projectiles such as protons and alpha particles.

3.2.2 Classical Derivation

Neils Bohr in 1913 [18, 19, 84] used a classical approach dependent on the impact-parameter b between the charged projectile's path and the nucleus of the target to evaluate the projectile's energy loss. Motivated by Rutherford's experiments on the scattering of alpha particles which helped establish the existence of the nucleus [102], Bohr set out using both classical electrodynamics and perturbation theory for distant collisions, leading to an adiabatic limit of energy transfer determined by electron orbital frequencies as well as a precise description of close collisions. Thus, he went beyond the Rutherford approach by accounting for atomic binding of the electron. Here are the main steps with emphasis on the mathematical aspects of the derivations[4]. The heavy projectile with mass M_1 and charge $Z_1 e$ moves at a velocity v, passes near an electron of the target with mass $m_e << M_1$ and charge e, at an impact parameter b (see Fig. 3.2). The target has atoms with atomic number Z_2. The parallel momentum transfer to a single electron averages to zero because of symmetry. The electron receives a transverse momentum impulse Δp given by

$$\Delta p = \int_{-\infty}^{\infty} e\mathbf{E}_T(t)dt = \int_{-\infty}^{\infty} e\mathbf{E}_T \frac{dt}{dx}dx \qquad (3.18)$$

$$= \int_{-\infty}^{\infty} e\mathbf{E}_T \frac{dx}{v} = \int_{-\infty}^{\infty} \frac{Z_1 e^2}{(x^2+b^2)} \times \frac{b}{\sqrt{x^2+b^2}} \frac{1}{v}dx \qquad (3.19)$$

$$= \frac{Z_1 e^2 b}{v}\left[\frac{x}{b^2\sqrt{x^2+b^2}}\right]_{-\infty}^{\infty} = \frac{2Z_1 e^2}{bv}, \qquad (3.20)$$

where $\mathbf{E}_T$ is the transverse electric field. The energy transferred ΔE is easily obtained as:

$$\Delta E = \frac{(\Delta p)^2}{2m_e} = \frac{2Z_1^2 e^4}{m_e v^2}\left(\frac{1}{b^2}\right). \qquad (3.21)$$

Eq. 3.21 assumes that relative to b the electron does not move much. Integrating the transferred energy ΔE in Eq. 3.21 over all possible values of the impact parameter b gives the energy lost per target atom simply as:

$$2\pi Z_2 \int \Delta E(b)bdb = 4\pi Z_2 \frac{2Z_1^2 e^4}{m_e v^2} \int_0^{\infty} \frac{1}{b^2}bdb. \qquad (3.22)$$

In Eq. 3.22, the integral diverges as $b \to 0$ and becomes undefined as $b \to \infty$. Bohr obtained a relatively precise b_{min} and a b_{max} based on adiabatic limit of energy transfer with respect to bound electrons orbiting

---

[4]For an alternative method of classical derivation consistent with Bohr's and starting from the Rutherford differential scattering cross section, see Jackson, Classical Electrodynamics, 3rd Ed, 2001 [53]. Similar approaches have been presented elsewhere [84]

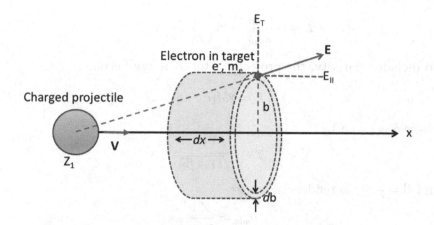

Figure 3.2 Schematic for Bohr's derivation of energy loss. The charged projectile approaches and interacts with an electron of the target. The Coulombic interaction leads to momentum transfer and energy loss. The direction of the electric field shown is that of a positively charged projectile.

at frequency ω. To get b_{min}, the closest distance of approach occurs with maximum energy transfer and must happen during a head-on collision. Rutherford already worked this out for two-particle elastic scattering [102] giving

$$b_{min} \approx \frac{Z_1 e^2}{m_e v^2}.$$

For b_{max}, no energy is transferred when the interaction is much longer than the orbital electron frequency, giving

$$b_{max} \approx \frac{v}{\omega}.$$

Evaluating Eq. 3.22 using these values of b_{min} and b_{max} gives Bohr's classically derived formula for energy loss of fast heavy charged projectiles in matter:

$$-\frac{dE}{dx} = \frac{4\pi Z_1^2 Z_2 e^4}{m_e v^2} \ln\left(\frac{m_e v^3}{Z_1 e^2 \omega}\right). \tag{3.23}$$

Thus, the stopping logarithm in the Bohr's formula is

$$L = \ln \frac{b_{max}}{b_{min}} = \ln \left(\frac{m_e v^3}{Z_1 e^2 \omega} \right). \tag{3.24}$$

To include relativity, the projectile's total energy becomes

$$E = \gamma M_1 c^2 \tag{3.25}$$

where γ is given by

$$\gamma = \frac{1}{\sqrt{(1 - \beta^2)}}$$

and $\beta = \frac{v}{c}$. This renders

$$b_{min} \approx \frac{Z_1 e^2}{\gamma m_e v^2}$$

and

$$b_{max} \approx \frac{\gamma v}{\omega}.$$

The stopping logarithm or stopping number thus takes the form:

$$L = \ln \left(\frac{\gamma^2 m_e v^3}{Z_1 e^2 \omega} \right). \tag{3.26}$$

Obviously, Bohr's formula for energy loss with relativistic projectile is:

$$-\frac{dE}{dx} = \frac{4\pi Z_1^2 Z_2 e^4}{m_e v^2} \ln \left(\frac{\gamma^2 m_e v^3}{Z_1 e^2 \omega} \right). \tag{3.27}$$

Example 3.2 *From relativistic considerations, calculate β for protons with a kinetic energy of 10 MeV and evaluate if the speed is truly relativistic.*

Solution

From Eq. 3.25, the total relativistic energy is

$$E = \gamma M_1 c^2,$$

therefore, kinetic energy E_k is the difference between the total energy and the rest energy:

$$E_k = M_1 c^2 (\gamma - 1), \tag{3.28}$$

which, generalizing for any particle with rest mass m_o becomes:

$$E_k = m_0 c^2 (\gamma - 1).$$ (3.29)

For the 10 MeV proton, $M_1 = m_p$ is 938.3 MeV/c^2 and we have:

$$E_k = M_1 c^2 (\gamma - 1) = M_1 c^2 (\frac{1}{\sqrt{(1 - \beta^2)}} - 1).$$

Thus,

$$\beta^2 = \frac{v^2}{c^2} = 1 - \frac{1}{(1 + \frac{E_k}{M_1 c^2})^2}.$$ (3.30)

Substituting with values,

$$\beta^2 = 1 - \frac{1}{(1 + \frac{10}{938.3})^2} = 0.0209.$$

So $v = \beta \times c = 0.1448 \times c = 4.34 \times 10^7$ m/s. This speed is truly relativistic: the speed parameter is greater than 0.14, the usual rule of thumb!

Exercise 38

From relativistic considerations, what is the speed of protons that have a kinetic energy of 100 MeV?

Exercise 39

Calculate the electron density n_e of lead in units of electrons/g. Hint: Use Eq. 3.6. For Pb, $Z_2 = 82$ and $A_2 = 207.2$ g/mol.

3.3 BETHE'S QUANTUM MECHANICAL FORMULA

3.3.1 Derivation of Differential Cross Section

Bethe's paper [11, 13] is a masterpiece with 91 equations, not counting subequations. The original German version [11] has been well translated and so the curious reader is encouraged to read Hans Bethe's **Theory of the Passage of Fast Corpuscular Rays Through Matter, 1930** [13] in English. Bethe started out using as first principles the observable energy and momentum transfers between fast charged projectiles with atomic electrons of target materials. He developed the excitation cross-section in the so-called first Born approximation which means that incident projectile is represented as a plane wave while the scattered projectile is a slightly perturbed wave. Hence, he used first order perturbation theory. Bethe represented the Coulombic interaction as a Fourier integral over momentum transfer and thereby derived the differential cross section for excitation to the nth quantum state of the target atom. Taking q^5 as the magnitude of the momentum transfer to an electron (in units of $\hbar$) and Q as the energy of the electron having that momentum q, the Born differential cross section is:

$$d\sigma_n = \frac{2\pi Z_1^2 e^4}{m_e v^2} \mid F_n(q) \mid^2 \frac{dQ}{Q^2}, \tag{3.31}$$

where

$$F_n(q) = \sum_j \langle n | e^{2\pi i h \mathbf{q} \cdot r_j} | 0 \rangle$$

is the so-called inelastic form factor which is the sum over all the coordinates of the target electron with the matrix elements of the momentum transfer operator $\mathbf{q}$ between the ground state and the nth excited state. The inelastic form factor is related to the transition probabilities through the c_j^1 of first order perturbation theory. Recall from Eq. 3.13 that

$$P_j = \mid c_j^{(1)}(\infty) \mid^2.$$

Hence, the $c_j^{(1)}$ are proportional to the $F_n(q)$. Bethe introduced the dipolar approximation to enable the summation for the exponential in $F_n(q)$, that is,

$$e^{2\pi i h \mathbf{q} \cdot r_j} \approx 1 + 2\pi i h \mathbf{q} \cdot r_j.$$

To obtain the total energy loss, Bethe used Eq. 3.31 thus:

$$-\frac{dE}{dx} = N \sum_n \int_{Q_{min}}^{Q_{max}} E_n d\sigma_n. \tag{3.32}$$

In Eq. 3.32, E_n is the energy of transition and N is the atomic number density. Q_{min} and Q_{max} denote the minimum and maximum values of Q with respect to specified excitation energy. Bethe divided the range of Q into large ($Q > Q_0$) momentum transfers and small values ($Q < Q_0$) momentum transfers. We have $Q < Q_0$, due to soft or glancing or distant collisions and $Q > Q_0$ due to hard or close or knock-on collisions. More

---

[5]This notation using q for momentum is due to Bethe himself though one would have preferred p for momentum, even as a transformed quantity.

explicitly, distant or soft collisions imply small momentum transfer and therefore small energy transfer. Close or hard collisions imply large momentum transfer and therefore large energy transfer. For soft collisions with $Q < Q_0$, the dipole approximation is valid and the $F_n(q)$ (which are proportional to the $c_j^{(1)}$) are given by:

$$| F_n(q) |^2 \longrightarrow Q \frac{f_n}{E_n}, \qquad (3.33)$$

where f_n is the corresponding dipole oscillator strength. That is, if $E_n = \varepsilon_j - \varepsilon_0$, then $f_n = f_j - f_0$. The contribution of soft or distant collisions to energy loss can thus be written as:

$$-\frac{dE}{dx}_{dist} = \frac{2\pi Z_1^2 e^4 N}{m_e v^2} \sum_n f_n \ln \frac{Q_0}{(E_n^2/2m_e v^2)}. \qquad (3.34)$$

Of course, Q_0 is the upper limit of Q for distant or soft collisions and Eq. 3.34 has

$$Q_{min} = \frac{E_n^2}{2m_e v^2}.$$

For large momentum transfers involved in hard or close collisions, Bethe proved the sum rule of Thomas-Reiche-Kuhn, with $\sum_n f_n = Z_2$ so that

$$\sum_n E_n | F_n(q) |^2 = Z_2 Q. \qquad (3.35)$$

The contribution of hard or close collisions is then calculated using:

$$-\frac{dE}{dx}_{close} = \frac{2\pi Z_1^2 e^4 N}{m_e v^2} Z_2 \ln \frac{2m_e v^2}{Q_0}. \qquad (3.36)$$

Notice that in Eq. 3.36, Q_0 is now the lower limit for hard or close collisions and the corresponding maximum is the kinematic maximum, $2m_e v^2$. This kinematic maximum itself is obtained by considering the maximum transfer energy T_{max} for a charged projectile interacting with an electron ($m_1 >> m_e$), using the effective mass:

$$T_{max} = \frac{4m_1 m_e}{(m_1 + m_e)^2} \frac{m_1 v^2}{2} \approx 2m_e v^2. \qquad (3.37)$$

To present the formula in terms of measurable quantities, and also in a format comparable with Bohr's and other previous works, Bethe defined the **mean excitation potential** of the target atom, I thus:

$$\ln I = \frac{1}{Z_2} \sum_n f_n \ln E_n. \qquad (3.38)$$

So I is the mean excitation energy of the target. Succinctly, for Bethe,

$$-\frac{dE}{dx} = N\sum_n \int_{Q_{min}}^{Q_{max}} E_n d\sigma_n = -\frac{dE}{dx}_{dist} - \frac{dE}{dx}_{close} \qquad (3.39)$$

$$= N\sum_n \int_{(E_n^2/2m_e v^2)}^{Q_0} E_n d\sigma_n + N\sum_n \int_{Q_0}^{2m_e v^2} E_n d\sigma_n. \qquad (3.40)$$

Combining Eq. 3.34 and Eq. 3.36, we have, finally,

$$\boxed{-\frac{dE}{dx} = \frac{4\pi Z_1^2 e^4 N Z_2}{m_e v^2}\ln\frac{2m_e v^2}{I}.} \qquad (3.41)$$

Bethe compared his result (Eq. 3.41) with Bohr's formula (Eq. 3.23) and noted as we can see here that the argument of the logarithm in Bethe's contains v^2 instead of the v^3 in Bohr's. They both have the same kinematic factors but differ in the logarithm number or stopping number.

3.3.2 Validity Conditions and Relativistic Corrections

Throughout the derivation of the Bethe formula for energy losses, we assumed that $m_1 \gg m_e$. Thus, energy loss is independent of the mass of the projectile and we do not see m_1 in the energy loss formulae. Also, we assumed that $v \gg v_0$, that is, the velocity of the projectile is very large compared to the velocities of atomic electrons so that $m_e v^2 \gg \hbar\omega_0$ but $v < c$. Before delving into relativistic corrections, we note that for $v \approx v_0$, the first order Born approximation used for the derivation of the Bethe formula is no longer valid. That is, first order perturbation method is no longer enough. Correction terms can already be envisaged in view of the values of v. In 1932, Bethe [12] extended his derivation to the relativistic case where

$$v \approx c.$$

With $m_1 \gg m_e$, kinematic maximum, $2m_e v^2$ now becomes

$$T_{max} = 2m_e \gamma^2 v^2 \qquad (3.42)$$

where

$$\gamma = \frac{1}{\sqrt{(1-\beta^2)}}$$

and $\beta = \frac{v_1}{c} \approx 1$. Anecdotally, the full expression for T_{max} in the relativistic limit before the restriction imposed by $m_1 \gg m_e$, is

$$T_{max} = \frac{2m_e c^2 \beta^2 \gamma^2}{(1 + 2\gamma m_e/m_1 + (m_e/m_1)^2)}. \tag{3.43}$$

These modifications in the velocity, engendering modifications in the momentum and energy lead to the following relativistic Bethe energy loss formula:

$$-\frac{dE}{dx} = \frac{4\pi Z_1^2 e^4 N Z_2}{m_e \gamma_1^2 v^2} \ln \frac{2m_e \gamma_1^2 v^2}{I}. \tag{3.44}$$

For easy back-of-envelope calculations and intuitive estimation, we note that

$$\frac{4\pi Z_1^2 e^4 N Z_2}{m_e \gamma_1^2 v^2} = \frac{4\pi Z_1^2 e^4 N Z_2}{m_e v^2} (1 - \beta^2)$$

and

$$\ln \frac{2m_e \gamma_1^2 v^2}{I} = \ln \frac{2m_e v^2}{I(1 - \beta^2)},$$

hence, we recast the relativistic Bethe energy loss formula thus:

$$-\frac{dE}{dx} = \frac{4\pi Z_1^2 e^4 N Z_2}{m_e v^2} (1 - \beta^2) \ln \frac{2m_e v^2}{I(1 - \beta^2)}, \tag{3.45}$$

which can be also rendered neatly but with the minor rearrangement of the Stopping number:

$$-\frac{dE}{dx} = \frac{4\pi Z_1^2 e^4 N Z_2}{m_e v^2} \times \left[\ln \left(\frac{2m_e v^2}{I(1 - \beta^2)} \right) - \beta^2 \right]. \tag{3.46}$$

Weaver and Westphal [123] have in recent times integrated the work of various authors over decades and compared results of calculations and extrapolations with new improved measurements especially using Uranium ions. They found interesting agreements. For easy comparison with data tables in the literature and for use in calculations as well as checking of experimental results, we note that Eq. 3.46 is consistent with the Bethe formula for **electronic or collisional mass stopping power** found in ICRU 37 [8], ICRU 49 [7] and ICRU 90 [108] and in the widely used online software platforms for calculations of mass stopping power such as ESTAR: Electronic Stopping Power and Range; PSTAR: Proton Stopping Power and Range and ASTAR: Alpha Particle Stopping Power and

Range all at the National Institute of Standards and Technology (NIST) website, https://physics.nist.gov/PhysRefData/Star/Text/intro.html:

$$\boxed{\frac{1}{\rho}\frac{dE}{dx} = \frac{1}{\rho}S_{col} = \frac{4\pi r_e^2 m_e c^2}{u}\frac{Z_1^2}{\beta^2}\frac{Z_2}{A_2} \times \left[\ln\left(\frac{2m_e c^2 \beta^2}{I(1-\beta^2)}\right) - \beta^2\right].} \quad (3.47)$$

In Eq. 3.47, note the replacement of v^2 in Eq. 3.46 with $\beta^2 c^2$, the use of r_e and so on. In anticipation of the notation in Eq. 3.58, we already stated the quantities found here back in Eq. 3.1 and Eq. 3.2. NZ_2 in Eq. 3.47 is the electron density in units of electron per g (Eq. 3.6):

$$n_e = NZ_2 = \frac{N_A Z_2}{A_2} \quad (3.48)$$

Thus,

$$N = \frac{N_A}{A_2}$$

which is the number of atoms or nucleons per gram, as used in the expression for the electron density of the absorber (Eq. 3.6), is also given by

$$N = \frac{N_A}{M_A} = \frac{1}{uA}$$

where M_A is the molar mass in g mol^{-1}, A is the relative molecular mass (or relative atomic mass) and u is the atomic mass unit ($\frac{1}{12}$ the mass of an atom of Carbon-12 isotope). To clarify further, the value of u is $1.6605402 \times 10^{-24}$g. Care must be taken while invoking n_e as the electron density because it can be stated as number of electrons per mass (eg, per g, as we have done above) or number of electrons per volume N_e as shown in Eq. 3.49

$$N_e = \rho N_A \frac{Z_2}{A_2}. \quad (3.49)$$

Furthermore, Eq. 3.47 is an experimentally convenient form of Bethe formula for collisional mass stopping power because $\frac{4\pi r_e^2 m_e c^2}{u}$ is a constant. $\frac{Z_2}{A_2}$ is also constant for many materials. Therefore, collisional mass stopping power depends largely on the charge of the charged projectile and its velocity with respect to the speed of light β. Additionally, since the results of experimental work often express the energy loss of charged particles in terms of the stopping cross section $\sigma_{Stopping}$ (in units of 10^{-15}

eV cm^2, we state next the relationship between stopping cross section and mass collision stopping power (in MeV $g^{-1}cm^2$):

$$\sigma_{Stopping} = 10^{21}\frac{M_A}{N_A}\frac{1}{\rho}S_{col} = 10^{21}\frac{M_A}{N_A}\frac{1}{\rho}S_{el}. \tag{3.50}$$

Example 3.3 *Express the Bethe formula in terms of the maximum kinetic energy T_{max} transferred to a single electron by the charged projectile.*

Solution

$$-\frac{dE}{dx} = 4\pi N_A r_e^2 m_e c^2 Z_1^2 \frac{Z_2}{A_2}\frac{\rho}{\beta^2}\left[\frac{1}{2}\ln\left(\frac{2m_e c^2 \gamma^2 \beta^2 T_{max}}{I^2}\right) - \beta^2\right]. \tag{3.51}$$

Note that the inclusion of T_{max} in the stopping logarithm engendered the squaring of the ionization potential, I.

3.4 BLOCH AND OTHER EXTENSIONS OF BETHE'S FORMULA

3.4.1 Summary of Corrections and Extensions

Arising from first order perturbation theory (see Appendix A.3), Bethe's formula has Z_1^2 terms only. Extensions and corrections were made by several others [1, 17, 32, 38, 69] to improve the formula. There are very good reviews, historical accounts and syntheses of the century-long development of energy loss formulas [7, 32, 128]. Among these formulas are those based on using higher powers of the projectile charge Z_1, which means using higher order perturbation methods instead of just first order perturbation. Walter H. Barkas and Hans Henrik Andersen made extensions proportional to Z_1^3 [128]. This accounts for the so-called Barkas-Andersen-effect which is the dependence of energy loss on projectile charge at very low energies. Although the Barkas-Andersen effect adds an extra Z_1 term to the stopping logarithm, it was called Z_1^3 correction because of combining Z_1 with the Z_1^2 in the Bethe formula. The Felix Bloch correction [17] corresponds to Z_1^4 and accounts for the low energy region or low velocity region where first order Born approximation used in the Bethe formula no longer holds. The Bloch correction was derived by Bloch in an investigation of the similarities and differences between classical and quantum mechanical range-energy calculations [17]. The Lindhard-Sorensen correction [69] is a modern replacement for the corrections of Bloch. A group of corrections with similar objectives and trend of development, known as the BMA group (Bloch, Mott and Ahlen) [1, 17] is common in the literature since it produces more precise stopping power calculations compared to separate corrections by Bloch, Mott and Ahlen. The Mott correction is based on Mott scattering. It addresses the scattering of electrons off nuclei with high charges. The Ahlen correction extends the Bloch correction to both high charge and energy [1]. The Lindhard-Sorensen correction [69] recovers the Bloch correction in the low-energy limit, while also incorporating Mott scattering in a relativistically correct manner [69]. Since the atomic electrons of the target material traversed are not stationary, there is need for **shell correction** which account for the fact that the projectile velocity v is not necessarily large compared to the orbital velocity of the target electrons v_0. Hence, the shell correction is designated $\frac{C(\beta)}{Z_2}$. Furthermore, Enrico Fermi [38] provided what is termed **density-effect correction** and this accounts for the reduction of the energy loss due to the polarization of the target medium. It is designated $\frac{\delta}{2}$. A typical plot of the Bethe formula for collisional or electronic mass stopping power (along with nuclear and total stopping power) is shown in Fig. 3.3. The features of this plot are discussed below.

3.4.2 Bloch's Correction and Extensions: Mathematical

Felix Bloch [17] compared Bohr's result with Bethe's result and undertook some slightly different mathematical approaches and we are concerned with the mathematical physics of nuclear experiments in this text. Without extensive history, we outline the steps Bloch took. Just as Bethe, Bloch considered charged projectiles with $v >> v_0$, that is, those with velocity very large compared to the velocities of atomic electrons so that $m_e v^2 >> \hbar \omega_0$. He then analyzed close or hard collisions within a Bohr-like cylindrical confinement without assuming the electrons to be plane waves. Thus, he allowed for interfering momentum components during the interaction. This also means that he used neither Born approximation nor the first order perturbation on which Born approximation relies. Within Bloch's cylinder large momentum transfers would be similar to Bohr's derivation and small transfers would agree with Bethe's plane wave approximation or Bohr approximation, enabling Bloch's result to retain the quantized energy/momentum approach of Bethe. For

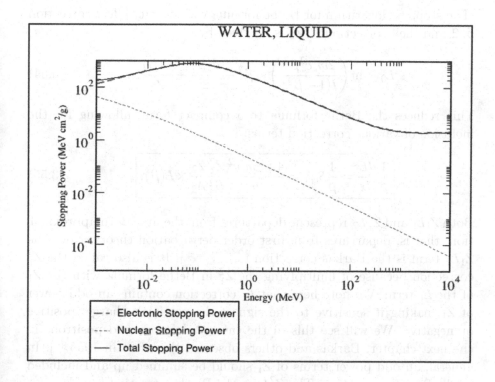

Figure 3.3 Mass Stopping Power of Water for Protons using data from PSTAR: Proton Stopping Power and Range (see National Institute of Standards and Technology (NIST) website, https:// physics.nist.gov/PhysRefData/Star/Text/PSTAR.html). The calculations were done using Bethe formula with corrections. Nuclear collisional loss is negligible. Electronic collisional loss is depicted by the Electron Stopping Power.

mathematical convenience, we express the stopping logarithm or stopping number as a sum of three terms [7,69], serving to express fine details including correction terms:

$$L(\beta) = L_0(\beta) + Z_1 L_1(\beta) + Z_1^2 L_2(\beta). \qquad (3.52)$$

Without further ado, $L(\beta)$, the stopping number per electron, is now conventionally [7,69,123] expressed as:

$$L(\beta) \equiv \ln\left(\frac{2m_e c^2 \beta^2}{I(1-\beta^2)}\right) - \beta^2 - \frac{\delta(\beta)}{2} - \frac{C(\beta)}{Z_2} + Z_1 L_1(\beta) + Z_1^2 L_2(\beta). \qquad (3.53)$$

The stopping logarithm for Bethe formula with density effect correction $\delta/2$ and shell correction C/Z_2 is L_0 given by:

$$L_0 = \ln\left(\frac{2m_ec^2\beta^2}{I(1-\beta^2)}\right) - \beta^2 - \frac{\delta(\beta)}{2} - \frac{C(\beta)}{Z_2}. \tag{3.54}$$

This reduces the Bethe formula to a compact form, allowing for the inclusion of various correction terms:

$$\boxed{\frac{1}{\rho}\frac{dE}{dx} = \frac{1}{\rho}S_{col} = \frac{4\pi r_e^2 m_e c^2}{u}\frac{Z_1^2}{\beta^2}\frac{Z_2}{A_2} \times L(\beta).} \tag{3.55}$$

Both Z_1L_1 and $Z_1^2L_2$ represent departures from the first Born approximation, that is, departures from first order perturbation theory. Now, the Z_1L_1 term is the Barkas correction [7, 108, 128]. It is also called the Z_1^3 correction because of multiplying the Z_1^2 in Bethe formula with the Z_1 of the L_1 term. We note here that L_1 correction contains an odd power of Z_1 making it sensitive to the sign of the particle's charge: positive or negative. We will see this in the case of electrons and positrons in the next chapter. Barkas and others observed the effect in 1955 [4]. In general, all odd power terms of Z_1 should be summed up and included in the Barkas correction. The $Z_1^2L_2$ is the Bloch correction [8, 17].

Bloch's extension of the Bethe formula was done using an impact parameter method (just as Bohr did) but with explicit quantum mechanical properties accounted for. Bloch expressed the correction term as:

$$Z_1^2L_2(\beta) = \psi(1) - Re[\psi(1+iy)] \tag{3.56}$$

where $y = Z_1\alpha/\beta$ and $Re\psi$ is the real part of the logarithmic derivation of the gamma function [8]. A transformation of Eq. 3.56 based on the properties of the gamma function [55] yields:

$$Z_1^2L_2(\beta) = -y^2 \sum_{n=1}^{\infty}[n(n^2+y^2)]^{-1}. \tag{3.57}$$

The Bloch correction is negligible when y is very small. For $y \gg 1$, $Z_1^2L_2$ approaches $-0.577 - \ln y$ and when Bloch added this value to the stopping number in the Bethe formula Eq. 3.47, he obtained the same value as the Bohr formula Eq. 3.27. In many calculations, the Z_1L_1 and $Z_1^2L_2$ corrections are not included and so it is appropriate to just stay with the designation, Bethe's formula (and not Bethe-Bloch) for collisional mass stopping power which should mean:

$$\frac{1}{\rho}\frac{dE}{dx} = \frac{1}{\rho}S_{col} = \frac{4\pi r_e^2 m_e c^2}{u}\frac{Z_1^2}{\beta^2}\frac{Z_2}{A_2} \times \left[\ln\left(\frac{2m_e c^2 \beta^2}{I(1-\beta^2)}\right) - \beta^2\right]. \quad (3.58)$$

As is often the case, when the density effect correction and the shell correction are added, we get the Bethe formula for collisional mass stopping power with corrections, which should mean:

$$\frac{1}{\rho}\frac{dE}{dx} = \frac{1}{\rho}S_{col}$$

$$= \frac{4\pi r_e^2 m_e c^2}{u}\frac{Z_1^2}{\beta^2}\frac{Z_2}{A_2} \times \left[\ln\left(\frac{2m_e c^2 \beta^2}{I(1-\beta^2)}\right) - \beta^2 - \frac{\delta(\beta)}{2} - \frac{C(\beta)}{Z_2}\right].$$

$$(3.59)$$

We can also express this most general form of the Bethe formula for collisional mass stopping power with corrections using N_A, C_1 and other symbols earlier used in Eq. 3.2, to facilitate numerical calculations:

$$\frac{1}{\rho}\frac{dE}{dx} = \frac{1}{\rho}S_{col}$$

$$= (4\pi r_e^2 m_e c^2)(\frac{Z_1^2}{\beta^2})(\frac{N_A Z_2}{A_2})\left[\ln\left(\frac{2m_e c^2 \beta^2}{I(1-\beta^2)}\right) - \beta^2 - \frac{\delta(\beta)}{2} - \frac{C(\beta)}{Z_2}\right],$$

$$(3.60)$$

which becomes

$$\frac{1}{\rho}\frac{dE}{dx} = \frac{1}{\rho}S_{col} = (C_1)(\frac{Z_1^2}{\beta^2})(\frac{N_A Z_2}{A_2}) \times L_0 \quad (3.61)$$

where

$$C_1 = 4\pi r_e^2 m_e c^2,$$

and

$$L_0 = \left[\ln\left(\frac{2m_e c^2 \beta^2}{I(1-\beta^2)}\right) - \beta^2 - \frac{\delta(\beta)}{2} - \frac{C(\beta)}{Z_2}\right].$$

Example 3.4 *How is the mean ionization potential I required for evaluating S or $\frac{1}{\rho}S_{col}$ calculated for elements and compound? State the Bragg additivity rule.*

Solution

For elements, I can be measured or estimated from various empirical formulas. For compounds, we use the Bragg additivity rule

$$\ln I = \frac{\sum_i N_i Z_i \ln I_i}{\sum_i N_i Z_i} \tag{3.62}$$

where N_i, Z_i indicate the number of atoms or fraction of the individual elements making the compound or mixture and the atomic number of the individual components, respectively.

Example 3.5 *There are several empirical rules for estimating the mean ionization potentials of elements. Given the following approximate rules,*

$$I(H) \approx 19\ eV \tag{3.63a}$$

$$I(2 \leq Z \leq 13) \approx 11.2 + 11.7Z \tag{3.63b}$$

$$I(Z > 13) \approx 52.8 + 8.71Z \tag{3.63c}$$

calculate the mean ionization for water.

Solution

Using the Bragg additivity rule (Eq. 3.62) and the empirical rules in Eq. 3.63a for hydrogen and 3.63b for oxygen, we have

$$I(H) \approx 19\ eV,$$

$$I(O) \approx 11.2 + 11.7(8) = 104.8\ eV,$$

So,

$$\sum_i N_i Z_i \ln I_i = 2 \times 1 \times \ln 19 + 1 \times 8 \times \ln 104.8 = 43.11$$

and

$$\sum_i N_i Z_i = 2 \times 1 + 1 \times 8 = 10.$$

Therefore,

$$\ln I = \frac{\sum_i N_i Z_i \ln I_i}{\sum_i N_i Z_i} = \frac{43.11}{10} = 4.311$$

Finally, $I = e^{\ln I} = 74.48 \approx 75 \, \text{eV}.$

Example 3.6 *Using the Bethe equation without corrections for density effect and without shell corrections, calculate the collisional mass stopping power of water for 10 MeV protons.*

Solution

We will use Eq. 3.60 in its condensed form (Eq. 3.61)

$$\frac{1}{\rho}\frac{dE}{dx} = \frac{1}{\rho}S_{col} = (C_1)\left(\frac{Z_1^2}{\beta^2}\right)\left(\frac{N_A Z_2}{A_2}\right) \times L_0$$

where

$$C_1 = 4\pi r_e^2 m_e c^2 = 5.099 \times 10^{-25} \; MeV.cm^2.$$

as we already calculated in section 1 above. Also we already found in the example above that for a 10 MeV proton,

$$\beta^2 = 0.0209.$$

Also calculated already are the electron density of 1 g of water:

$$n_e = \left(\frac{1\,mole}{18.0153\,g}\right)(6.022 \times 10^{23} \, \frac{molecules}{mole})(10\,\frac{electrons}{molecule})$$

$$= 3.343 \times 10^{23} \, \frac{electrons}{g}.$$

and the mean ionization potential of water $I = 75\,eV$. Recall that $n_e = \frac{N_A Z_2}{A_2}$. For protons $Z_1 = 1$. For the stopping logarithm without density effect and shell corrections, we have

$$L_0 = \left[\ln\left(\frac{2m_e c^2 \beta^2}{I(1-\beta^2)}\right) - \beta^2\right]$$

$$= \left[\ln\left(\frac{2 \times 0.511 \times 10^6 \, eV \times 0.0209}{75 \, eV(1 - 0.0209)}\right) - 0.0209\right] = 5.652$$

Hence, for

$$\frac{1}{\rho}S_{col} = (C_1)(\frac{Z_1^2}{\beta^2})(\frac{N_A Z_2}{A_2}) \times \left[\ln\left(\frac{2m_e c^2 \beta^2}{I(1-\beta^2)}\right) - \beta^2\right]. \tag{3.64}$$

we substitute the values to obtain:

$$\frac{1}{\rho}S_{col} = (5.099 \times 10^{-25} MeV \cdot cm^2)(\frac{1^2}{0.0209})(3.343 \times 10^{23} \frac{electrons}{g}) \times 5.652$$

$$= 46.1 \ MeV \cdot cm^2/g$$

Finally, the collisional or electronic mass stopping power of water for 10 MeV protons is:

$$\frac{1}{\rho}S_{col} = 46.1 \ MeV \cdot cm^2/g.$$

Example 3.7 *Calculate the velocity of a deuteron for which the mass collisional stopping power of water is the same as for a 10 MeV proton, as found in the example above. Hint: For a heavy charged projectile to engender in an absorber the same mass collision stopping power as a proton, the projectile must have the same velocity ratio β as well as the same charge number Z_1.*

Solution

In the Bethe formula, Eq. 3.64, we neatly arranged the terms to show that C_1 is constant, $\frac{Z_1^2}{\beta^2}$ and the stopping logarithm depend on the properties of the charged projectile while the electron density $\frac{N_A Z_2}{A_2}$, depends on the absorber. Now, for both proton and deuteron, the charge number $Z_1 = 1$. Thus, they must both have the same velocity and same velocity ratio β^2 to cause the same mass collision stopping power in an absorber. As shown earlier, $v = 4.34 \times 10^7 m/s$.

Exercise 40

(a) A beam of 100 MeV protons is incident on water. Calculate the speed ratio parameter β^2 (b) What is the mass collisional stopping power of water for the 100 MeV protons? Use Bethe's formula without density effect and shell corrections, as in Eq. 3.64.

Exercise 41

(a) A beam of 1 MeV protons is incident on water. Calculate the speed ratio parameter β^2 (b) What is the mass collisional stopping power of water for the 1 MeV protons? Use Bethe's formula without density effect and shell corrections, as in Eq. 3.64.

Example 3.8 *Using data from PSTAR: Proton Stopping Power and Range, NIST, plot a graph of the Bethe formula with corrections for protons in water. Use the stopping power in $MeV\,g^{-1}cm^2$ instead of the energy loss. Note and explain at least two key regions of the plot.*

Solution

The plot is shown in Fig. 3.3. The diligent reader will benefit from going to the website to make the plot. Key features of the plot include:

1. *The electronic mass stopping power rises at low energies (below 10^{-2} MeV up to about 10^{-1} MeV.*

2. *There is a logarithmic drop in electronic mass stopping power for energies between 10^{-1} MeV and 10^3 MeV.*

3. *At higher energies beyond 10^3 MeV, the electronic stopping power seemingly flattens out.*

As mentioned in Chapter two, the Landau distribution is sometimes used to characterize the measured heavy charged particle decay spectra, especially those of alpha particles. This is specifically the case for thin absorbers. For thick absorbers, the energy distribution becomes more symmetric and more approximately Gaussian. Furthermore, the energy loss is a statistical process since the number of collisions and amount of energy loss vary from particle to particle. This variation leads to an asymmetric distribution. Collisions leading to small energy transfers are more probable than those with large energy transfers. There is a tail at the very high energy loss values since the collisions with such high energy transfers are relatively rare. In some of these high-energy collisions, high energy electrons called δ-rays are produced.

Exercise 42

Using the values of the standard numerical constants (see front material), what is the value of $\frac{4\pi r_e^2 m_e c^2}{u}$ found in the collisional energy loss formula?

Exercise 43

Since $\frac{Z}{A}$ shows up in the Bethe formula and in fundamental relations such as (i) the number of electrons per volume of an element

$$\frac{ZN_a}{V} = \frac{\rho ZNa}{m} = \rho N_A \frac{Z}{A};$$

(ii) the number of electrons per mass of an element

$$\frac{ZN_a}{m} = N_A \frac{Z}{A};$$

where N_a is the number of atoms per mass of the element, (a) what is $\frac{Z}{A}$ for all elements except hydrogen? (b) What is the trend of $\frac{Z}{A}$ from low to high Z elements, using He-4, Co-60 and U-235?

3.5 RANGE OF HEAVY CHARGED PARTICLES

Historically, ranges of charged particles in matter were measured before stopping power formulas were derived. William Henry Bragg in 1905 published (along with Kleeman) an influential work: "On the α particles of radium, and their loss of range in passing through various atoms and molecules" [21]. They measured the ranges in air of the alpha particles at specific pressures. Although the energy loss formula enables the calculation of the range of charged particles traveling in matter, the fact that the range is experimentally measurable enables the testing of the stopping power theory. After such testing, the theory then gives its predictive power where the range cannot be directly measured, eg, in human tissue during teletherapy using protons and other heavy charged particles. When the energy loss is plotted as a function of the penetration depth we get the so-called *Bragg Curve*. Energy loss along the flight path causes the projectile to slow down and the Bethe formula predicts largest energy losses near the end of the track, leading to the *Bragg Peak* of the track. More intuitively, the energy loss of heavy charged particles due to collisions with atomic electrons of absorbers is gradual because they are much heavier than electrons and the particles continue in relatively straight paths because they are only slightly deflected by the electrons of the absorbers. Towards the end of its track, the charged projectile experiences charge exchange between the absorbing material, and picks up electrons until it becomes a neutral atom though the material receives

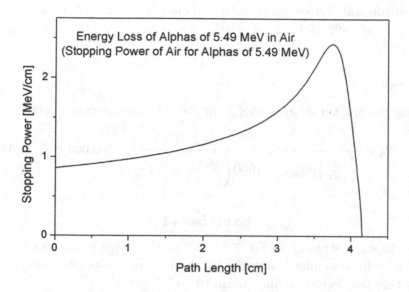

Figure 3.4 Bragg Curve for Alpha Particles in Air. The Energy-loss as a function of distance in air for typical alpha particle with energy 5.49 MeV emitted through radioactive decay. Graph was plotted by Helmut Paul and made available in public domain.

a net change in charge. Fig. 3.4 shows a typical Bragg curve for a 5.49 MeV alpha particle traveling through air. It is a plot of energy loss as a function of range in the medium. It is also described as Bragg pristine peak. A similar shape is obtained if the energy loss in MeV/cm is plotted against the kinetic energy (in MeV) of the particle in reverse.

An empirical quantity called *continuous slowing down approximation* (CSDA) is used to quantify the range. The CSDA range R_{CSDA} is given as:

$$R_{CSDA} = \int_0^{T_0} \left(\frac{dT}{\rho dx} \right)^{-1} dT, \tag{3.65}$$

where T_0 is the initial kinetic energy of the charged projectile. Since $\left(\frac{dT}{\rho dx} \right)$ has units MeV cm^2 g^{-1}, R_{CSDA} has units g/cm^2. To obtain range with dimension of length, one simply multiplies R_{CSDA} with the mass density of the material, ρ.

Example 3.9 *For an alpha particle of energy 5 MeV and average energy loss rate of 1000 MeV·cm² /g, what is the range in tissue? Take $\rho_{tissue} = 1\ g/cm^3$.*

Solution

Using Eq. 3.65 in its evaluated form for the given average energy,

$$R_\alpha = \frac{T_0}{\left(\frac{dT}{\rho dx}\right)\rho_{tissue}} = \frac{5\ MeV}{1000\left(\frac{MeV\cdot cm^2}{g}\right)\left(1\frac{g}{cm^3}\right)} = 0.005\ cm \quad (3.66)$$

Exercise 44

Use the same strategy in Eq. 3.66 to find the range of the same alpha particles in aluminium. Assume the same average specific energy loss and take the density of aluminium to be 2.7 g/cm³.

After determining the range of α particles in air, Bragg and Kleeman developed an empirical relationship to determine the range of other materials. The relationship is the so-called *Bragg-Kleeman rule* (see [118]) which is usually accurate to ±15%:

$$\frac{R_1}{R_2} \approx \frac{\rho_2}{\rho_1}\sqrt{\frac{A_1}{A_2}}, \quad (3.67)$$

where the subscripts 1 and 2 denote the quantities for the material of measured range R and material of desired range, respectively. A is usually atomic mass of the element or the the effective atomic mass of the compound or mixture. Atomic mass numbers can also be used consistently as A in Eq. 3.67. For clarity, Eq. 3.67 can also be stated thus:

$$R_1\rho_1\sqrt{A_2} \approx R_2\rho_2\sqrt{A_1} \quad (3.68)$$

Empirical and semi-empirical relations have been developed for the estimation of the range of frequently used heavy charged particles such as alpha particles and protons. For alpha particles in air, at 15°C and 760 mm Hg (1 atm), the range is

$$R_{air}[cm] = \begin{cases} 0.56\,Q_\alpha, & \text{if } Q_\alpha < 4 \text{ MeV.} \\ 1.24\,Q_\alpha - 2.62, & \text{if } 4 \le Q_\alpha < 8 \text{ MeV.} \end{cases} \quad (3.69)$$

A similar semi-empirical relation for the range in air R_{air} in cm of alpha particles at 0°C and 760 mm Hg, with energy Q_α ranging from 2 MeV to 8 MeV is approximately [23]:

$$R_{air}[cm] = 0.322 Q_\alpha^{\frac{3}{2}}, \quad \text{if } 2Q_\alpha \leq 8 \text{ MeV.} \tag{3.70}$$

One can see from the empirical relationships of Eq. 3.69 and Eq. 3.70 that the range of alpha particles is just a few centimeters in air. In fact, air is the most commonly used absorbing medium for specifying range-energy relationships of alpha particles, no wonder experiments in nuclear instruments courses usually involve the determination of the range of alpha particles in air. Furthermore, the effective atomic composition of biological tissue is similar to that of air. This makes it possible to estimate the range of alpha particles is in tissue using the Bragg-Kleeman rule of Eq. 3.67 and the empirical relations of Eq. 3.69. The result is:

$$R_{tiss} \approx R_{air} \frac{\rho_{air}}{\rho_{tiss}} \approx 1.293 \times 10^{-3} R_{air} \tag{3.71}$$

where we have used $\rho_{air} = 1.293 \times 10^{-3} \frac{g}{cm^3}$ at STP and $\rho_{tiss} \approx \rho_{water} = 1 \frac{g}{cm^3}$.

Example 3.10 *An alpha particle and a proton with the same energy enter a medium, say water. If the range of the proton is 10 cm, what is the range of the alpha particle? Take 938 MeV/c^2 as proton rest mass.*

Solution

Using Eq. 3.65

$$R_{CSDA} = \int_0^{T_0} \left(\frac{dT}{\rho dx} \right)^{-1} dT,$$

we see that range in length unit is

$$R_{CSDA} = \frac{E}{\frac{dE}{dx}}.$$

Bethe's formula for $\frac{dE}{dx}$ (Eq. 3.46) basically gives us:

$$\frac{dE}{dx} \propto \frac{z_1^2}{v^2} \tag{3.72}$$

and particle speed v is given by

$$v = \sqrt{\frac{2E}{m}}.$$

Since the medium is the same for both particles, z_2, ρ and A_2 in Bethe's formula are all the same. Thus,

$$\left(\frac{dE}{dx}\right)_{proton} \propto \frac{z_1^2}{v^2} = \frac{z_1^2}{\left(\frac{2E}{m}\right)} = \frac{z_1^2 m}{2E} = \frac{(1^2)(938 \text{ MeV}/c^2)}{2E} = \frac{938}{2E}.$$

Likewise,

$$\left(\frac{dE}{dx}\right)_{\alpha} \propto \frac{z_1^2}{v^2} = \frac{z_1^2}{\left(\frac{2E}{m}\right)} = \frac{z_1^2 m}{2E} = \frac{(2^2)(4 \times 938 \text{ MeV}/c^2)}{2E} = \frac{15008}{2E}.$$

We can now use their ratio of energy loss rates to obtain ratio of ranges:

$$\frac{\left(\frac{dE}{dx}\right)_{\alpha}}{\left(\frac{dE}{dx}\right)_{proton}} = \frac{\frac{15008}{2E}}{\frac{938}{2E}} = 16 \tag{3.73}$$

Since the α particle experiences 16 times more energy loss than the proton, the range of the proton will be 16 times greater. Hence

$$\frac{R_{\alpha}}{R_{proton}} = \frac{1}{16} \tag{3.74}$$

And in the given problem,

$$R_{\alpha} = \frac{R_{proton}}{16} = \frac{10 \text{ cm}}{16} = 0.625 \text{ cm}.$$

3.6 MEDICAL APPLICATIONS OF BRAGG PEAK

Since energy transfer of heavy charged particles to medium increases with decreasing momentum, the dose deposition increases as the charged particle slows down, dumping most of the energy towards the end of its path. When the medium is tissue, this longer range is advantageous for treating deep-seated tumors. If the energy loss of protons or other high energy projectiles, in MeV/cm, is plotted (on the y-axis) as a function of the kinetic energy in MeV, one obtains a curve similar to the pristine peak in Fig. 3.5, which is also similar to Fig. 3.4. For purposes of radiation therapy against cancer, the stopping power of a material is proportional to the dose or absorbed dose D deposited in the material. Fig. 3.5 shows the application of Bragg Peaks in medical physics where the absorbed dose is plotted as a function of range. The so-called spread

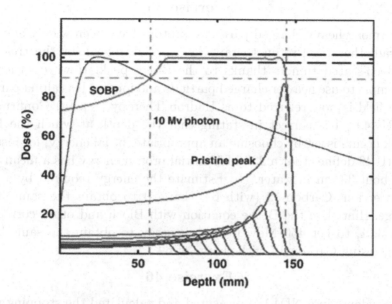

Figure 3.5 Percentage absorbed dose as a function of depth for a spread-out Bragg peak (SOBP) and its 12 constituent pristine Bragg peaks. The SOBP delivers higher and more uniform dose than a comparable 10 MV photon beam shown here. The SOBP dose distribution is produced by adding the contributions of individually modulated pristine Bragg peaks. In proton therapy, the clinical target volume is often larger than the width of a pristine Bragg peak. By appropriate modulation of the proton range and fluence of pristine peaks, the extent of the high-dose region can be widened to cover the target volume with a uniform dose. Adapted from [67].

out Bragg peak (SOBP) is obtained by stacking multiple depth dose curves of pristine peaks of different energies (various colors under the pristine peak). This enables a fairly uniform dose (SOBP in Fig. 3.5) to be applied to a tumor volume of given size or known size. Note that owing to their light mass, electrons and positrons undergo much more multiple scattering than protons, alpha particles and ions, and therefore do not exhibit the Bragg peak of heavy charged particles.

Exercise 45

Among heavy charged particles, protons have been widely employed in radiotherapy. Proton therapy has special appeal for the treatment of deep-seated tumors thanks to the Bragg peak. However, there are attempts to use heavier charged particles and ions such as alpha particles. The field is now referred to as Hadron Therapy [87]. (a) Confirm that ICRU 49 [7] is correct in stating that the depth at which the Bragg peak occurs is set by choosing an appropriate initial energy; for instance, ICRU 49 defines that a 230 MeV initial proton energy has a mean range of about 33 cm in water. (b) Estimate the energy required by a much heavier ion, Carbon-12 (with 6+ charges) to obtain the same 33 cm range. Hint: Use the Bethe equation with Bloch and other corrections, Eq. 3.61, to get 450 MeV/u as the energy to obtain the same 33 cm range using Carbon-12 with charge of 6+.

Exercise 46

Students in a NIM lab measured and calculated the stopping power and range of alpha particles in various copper foils of known thicknesses. Alpha particles from the decay of Po-210 (5.31 MeV) were detected using a surface barrier detector. A graph of the range of Po-210 alpha particle in copper was measured (see Fig. 3.6) to be $9.24 \pm 0.23 \, \text{mg/cm}^2$ which is in agreement with an expected value of $9.65; \text{mg/cm}^2$ according to Northcliffe and Schilling [86], but deviates from a $10.8 \, \text{mg/cm}^2$ published by NIST (ASTAR: Alpha Particle Stopping Power and Range, url: https://physics.nist.gov/PhysRefData/Star/Text/ASTAR.html). Speculate the reason(s) for the apparent disparity.

3.7   IDENTIFICATION OF PARTICLES AND OTHER APPLICATIONS

Contemporary high energy physics research uses massive detectors such as the ATLAS (A Torodial LHC ApparatuS) detector at the Large Hadron Collider at CERN to extract physical data of interactions between particles. The energy loss formula then enables the identification of particles from the interaction data and this is a major focus of high energy physics and particle physics in general. This is because the energy loss when plotted as a function of $\beta\gamma$ produces a universal curve that splits up for different particle masses, when taken as a function of energy or momentum. That is, a simultaneous measurement of energy

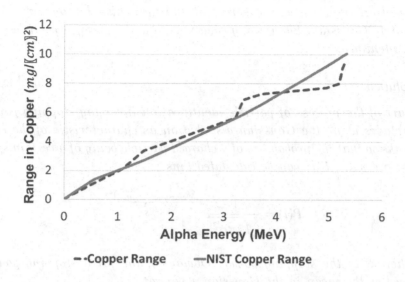

Figure 3.6 Range of alpha particles in copper of known thicknesses. Plots for measured range (dashed lines) and values from NIST data (solid line) are shown. Data was taken using a surface barrier detector with multichannel analyzer set for multiscaling, by Creighton University NIM students, Fall 2019, analyzed and tabulated by Oluyemi Bright Aboyewa (lab partners: John Higgins and Alec Peck.)

loss and momentum (in a magnetic field, for instance) enables particle identification. Although energy loss is independent of projectile mass, it is dependent on its momentum or energy via velocity and this is what enables particle identification. In general,

- range measurement can give energy of the particle E

- ionization measurement can give the velocity v

- delta-ray density measurement can give the charge number of the particle, Z_1

and various combinations of the parameters Z_1, E and v can used to uniquely fix the mass m of the particle, for identification. We briefly consider the mathematics employed in these and similar applications.

Example 3.11 *The energy spectra of various charged particles in a material are obtained by measurement and their distributions are approximately Gaussian. Show using equations how the charged particles might be identified.*

Solution

Part of the process of particle identification involving stopping power S, includes using the Gaussian distribution as characteristic of the energy peaks so that the probability of a charged particle being of a certain species $i = p, e, \pi, ...,$ $P(i)$ can be calculated thus

$$P(i) = \frac{1}{\sqrt{2\pi\sigma_S^2}} e^{-\frac{(S_{measured} - S_{fitted})^2}{2\sigma_S^2}} \tag{3.75}$$

where S is the stopping power. Obviously, the fitted stopping power is used as the mean in the Gaussian exponent.

Exercise 47

Collisional energy loss is proportional to the square of particle charge number and velocity:

$$\frac{dE}{dx} \propto \left(\frac{Z_1}{v}\right)^2 \tag{3.76}$$

First, rewrite this proportionality using the relativistic expression for velocity to obtain:

$$\frac{dE}{dx} \propto Z_1^2 \frac{(1 + \frac{m}{E})^2}{1 - (\frac{m}{E})^2}. \tag{3.77}$$

Then obtain the following proportionality relationship between energy loss and particle characteristics:

$$\Delta E \propto \left(\frac{2Z_1^2 m}{E}\right) \Delta x. \tag{3.78}$$

Hint: Expand the numerator of Eq. 3.77 ignoring all terms of the order of $(\frac{m}{E})^2$ and assume that $m \gg E/2$ since the kinetic energy is much less than projectile rest mass.

Example 3.12 *Give in outline how Eq. 3.78 can be used for charged particle identification, keeping in mind that the quantities involved can be independently measured using multiple detectors so as to enable the measurement of needed unknown quantities.*

Solution

1. *Since* $\Delta E \propto \frac{1}{E}$, *a plot of log E versus log* ΔE *will be linear, with a negative slope.*

2. *Using a fixed combination of mass and kinetic energy, energy loss* $\frac{\Delta E}{\Delta x}$ *then increases as* Z_1^2

3. *Using a fixed combination of* Z_1 *and E,* ΔE *increases directly as particle mass m*

Example 3.13 *Another method of particle identification involves the use of the parameter pβ (product of momentum and velocity ratio) which can be obtained from multiple scattering measurements. A plot of mean ionization energy I or stopping power, versus pβ gives a family of curves each unique to various charged particles. From stopping power and range measurements, the kinetic energy T of a particle is found to be 200 MeV and the momentum p = mγβ is found to be 490 MeV/c. (a) What is the mass of the particle? (b) What is the identity of the particle?*

Solution

Using natural units with $c = 1$,

$$T = (\gamma - 1)m = 200,$$

$$p = m\gamma\beta = 490$$

Since

$$\beta = \frac{\sqrt{\gamma^2 - 1}}{\gamma}, \tag{3.79}$$

$$p = m\gamma\frac{\sqrt{\gamma^2 - 1}}{\gamma} = 490,$$

and therefore

$$\frac{\sqrt{\gamma^2 - 1}}{\gamma - 1} = \frac{490}{200} = 2.45.$$

Rearranging,

$$\frac{\sqrt{\gamma+1}}{\gamma-1} = \frac{490}{200} = 2.45.$$

Solving for γ, we obtain $\gamma = 1.4$ Finally, the result of measurement,

$$T = (\gamma - 1)m = 200 \; MeV$$

is combined with the calculated $\gamma = 1.4$ to obtain the required mass as $m = 500 \; MeV/c^2$. In terms of electron rest mass, m_e, the mass of the unknown particle is:

$$m = \frac{500}{0.511} = 978.5 m_e.$$

This value is very close that that of a Kaon, that is, $966 m_e$.

3.8 PSTAR, ASTAR AND OTHER SOFTWARE PACKAGES

ESTAR, ASTAR and PSTAR stand out and these are available at the National Institute of Standards and Technology (NIST) website, https://physics.nist.gov/PhysRefData/Star/Text/intro.html. As noted already, ESTAR:Electronic Stopping Power and Range; PSTAR:Proton Stopping Power and Range; and ASTAR: Alpha Particle Stopping Power and Range all provide data for the calculation of mass stopping power and ranges of charged particles in matter. In addition to these, there is SRIM (Stopping and Range of Ions in Matter), a software package that has thousands of users around the world. It has been continuously upgraded since its introduction in 1985 by James Ziegler. A recent textbook, "SRIM-The Stopping and Range of Ions in Matter" describes in detail the fundamental physics of the software [129]. Details can easily be obtained from SRIM (url: http://www.srim.org). There is also TRIM which stands for Transport of Ions in Matter. TRIM is closely connected with SRIM. TRIM calculates 3D distributions of ions in up to 8 layers of selected materials as well certain kinetic phenomena associated with the ion's energy loss such as ionization and phonon production. Furthermore, there are codes for evaluating the various corrections to the Bethe formula: Bloch, Barkas and Shell corrections such as the BEST (BEthe STopping) developed by M. J. Berger and H. Bichsel [16] and used extensively in ICRU 49 [7]. The BEST includes the evaluation of the density-effect correction.

3.9 RADIATIVE LOSS VIA BREMSSTRAHLUNG FOR HEAVY CHARGED PARTICLES

High energy charged particles do undergo an additional energy loss due to bremsstrahlung, i.e. radiation of photons, in the Coulomb field of the atomic nuclei. We emphasize right away that bremsstrahlung is much stronger for lighter particles such as electrons and positrons [45] than for heavy charged particles such as protons, alpha particles, and heavy charged nuclei (fission fragments). Moreover, the effect can be neglected at particle energies below about 1 MeV, because the energy loss due to bremsstrahlung is very small. The average rate of radiative energy loss for high energy charged particles in general can be expressed thus:

$$-\frac{dE}{dx}\bigg|_{Brems} \approx 4\alpha N_A \frac{e^4}{m_1^2 c^4}\left(\ln\frac{183}{Z_2^{1/3}}\right)\frac{Z_2(Z_2+1)}{A_2}q_1^2 E. \tag{3.80}$$

where the subscript 2 refers to the properties of the material and subscript 1 refers to the properties of the incoming charged projectile, α is the fine structure constant, q is electric charge in units of e and E is the energy of the incoming projectile. The other symbols retain their previous definitions in this work. Eq. 3.80 is an approximation: a single closed form expression describing the radiative energy loss by charged particles for a large range of kinetic energies is not yet available. As an approximation, resulting from Bethe and Heitler's 1934 [14] work, before we note the outstanding dependencies in Eq. 3.80, let us briefly outline the steps taken to arrive at the Bethe-Heitler formula approximated in Eq. 3.80. Using perturbation theory and Born approximation (plane waves for electrons), Bethe-Heitler first derived a differential cross section for a fast moving charged projectile decelerated in the Coulomb field of a target material, with energy loss large enough to cause not only ionization but radiation of photons. The energy loss therefore entails radiative emission of photons which carry away a fraction of the kinetic energy of the incident projectiles. The Bethe-Heitler formula for the differential cross section in this process then gets corrected for various effects such as the screening field of the nucleus, the restrictions of the Born first approximation (so Coulomb waves replace plane waves for electrons and we get a Coulomb correction function), the polarization of the target material (dialectric suppression) and even the contribution of bremsstrahlung from atomic electrons. Seltzer and Berger [7, 8, 108, 109] have synthesized these correction theories nicely and the details are beyond the scope of this work. A few of these theories and corrections will be highlighted in the next chapter for electron-electron bremsstrahlung. Here, we note the outstanding dependencies of the Bethe-Heitler formula for high bremsstrahlung energy loss by energetic charged particles: (a) the dependence on the atomic number and atomic mass of the material is obvious:

$$-\frac{dE}{dx}_{Brems} \propto \frac{Z_2(Z_2+1)}{A_2};$$

(b) the dependence on the incident charged projectile is

$$-\frac{dE}{dx}_{Brems} \propto \frac{1}{m_1};$$

hence, light particles such as electrons, positrons and muons radiate more (c) the dependence on the energy of the incoming projectile:

$$-\frac{dE}{dx}_{Brems} \propto E;$$

and so the higher the energy, the greater the radiative losses via bremsstrahlung. We shall give more details in the treatment for electrons and positrons (Chapter 4).

Exercise 48

Assuming the mechanism of energy loss for electrons and protons in a material is the same (not exactly true: see Chapter 4), find the equivalent kinetic energy at which an electron would have the same rate of energy loss as a 10 MeV proton.

Exercise 49

Electron density influences many interactions of radiation and charged particles with matter. What is the expression for electron density in units of electrons/g, in terms of Avogadro's number N_A, atomic number Z and atomic mass number A?

Exercise 50

In a NIM experiment, you find the range of alpha particles from Po-210 to be 3.93 cm in air. Estimate the range of this Po-210 alpha particles in tissue. Hint: Use Bragg-Kleeman rule in the specific form worked out in Eq. 3.71 for tissue-air conversion.

Exercise 51

For a 5.31 MeV alpha particle emitted by Po-210, students in a NIM lab obtained a range of 0.0051 cm in mylar, a tissue equivalent material. What is the mean energy loss rate $\left(\frac{dE}{dx}\right)$ in tissue?

Exercise 52

(a) Calculate the mass stopping power of water for a 50 MeV proton using the Bethe mass collision stopping power equation without density effect or shell corrections, Eq. 3.64. Hint: Follow the procedure used in the calculation for 10 MeV proton in water. (b) Does your result agree with the data provided by NIST?

3.10 ANSWERS

Answer of exercise 38
Ans: $v = 0.428c = 1.284 \times 10^8$ m/s.

Answer of exercise 39
Ans: $n_e = 2.388 \times 10^{23}$ electrons/g.

Answer of exercise 40
Ans: (a) $\beta^2 = 0.183$; (b) 7.3 MeV.cm^2/g.

Answer of exercise 41
Ans: (a) $\beta^2 = 0.002128$; (b) 267.7 MeV.cm^2/g.

Answer of exercise 42
Ans: 0.307075 MeV.cm^2/g.

Answer of exercise 43
Solution: (a) Z/A is about 0.5; (b) Z/A. Actually, Z/A slowly decreases from 0.5 for low Z elements to 0.4 for high Z elements, for example, Z/A for He-4 = 0.5, for Co-60 =0.45 and for U-235 is 0.39.

Answer of exercise 44
Ans: Range of alphas in aluminum = 0.0019 cm.

Answer of exercise 45
Solution: 450 MeV/u to obtain the same 33 cm range using 12-Carbon-6+.

Answer of exercise 46

Solution: Stopping power correction.

Answer of exercise 47

Steps given in the text.

Answer of exercise 48

Solution. Set $dT/dx = (z^2 \times m/2E) = (dT/dx)_{electron} = (dT/dx)_{proton}$. Let $m_p = 1836m_e$. Cancel out the m_e. Solve for E. $1/2E = 91.8$. So $E = 0.005$ MeV $= 5.0$ keV.

Answer of exercise 49

Solution $N_A Z/A$.

Answer of exercise 50

Solution: $R_{tiss} = R_{air} \times \rho_{air}/\rho_{tiss} : 5.1 \times 10^{-3} cm$.

Answer of exercise 51

Solution: 5.3 MeV$/0.0051$cm $= 1.04 \times 10^3$ MeV/cm.

Answer of exercise 52

(a) solution: For water

$$N_e = 3.343 \times 10^{23} electron/g$$

$I_{water} = 74.4$ eV and β for 50 MeV proton is 0.314. So Collisional Stopping Power is 12.43 MeV.cm^2/g. (b) Yes. NIST gives 12.45 MeV.cm^2/g.

CHAPTER 4

Energy Loss of Electrons and Positrons through Matter

Although there are mechanisms of energy loss that are common to both light charged particles (this chapter) and heavy charged particles (Chapter 3), such as collisions with electrons involving energy transfer and Bremsstrahlung, we explore the mathematical physics used in reckoning these processes by focusing on the mechanisms dominant for each group of charged particles without implying that other mechanisms of energy losses are not possible. A summary of the possible mechanisms of energy losses for heavy charged particles, light charged particles, photons (treated in Chapter 5) and even neutral projectiles is presented in Fig. 4.1. Of course, properties of the projectiles such as their charge, mass, velocity and energy determine the relative predominance of the energy loss processes. In this chapter, we consider the energy loss processes for light charged particles, especially electrons and positrons which, in the context of radioactivity, are collectively known as beta particles. Beta particles or beta rays are emitted by certain fission fragments or by certain radioactive nuclei and constitute a class of ionizing radiation.

4.1 COLLISIONAL LOSS AND MODIFIED BETHE FORMULA

Here, we refer to electrons and positrons as light charged particles. They are also called Beta particles. They are among the light charged particles earlier excluded from the treatment in chapter three. Beta particles de-

DOI: 10.1201/9781003215622-4

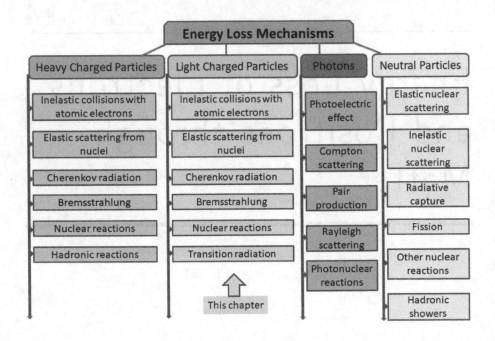

Figure 4.1 Energy Loss Mechanisms for Heavy Charged Particles, Light Charged Particles, Photons and Neutral Projectiles. Properties of the projectiles such as charge, mass, velocity and energy determine the relative predominance of the processes. This chapter deals with light charged particles.

posit energy in matter through two main processes, namely, collisional losses via excitation and ionization of atoms and radiative losses via Bremsstrahlung. Hence, the total energy loss for light charged particles is equal to the sum of both:

$$-\frac{dE}{dx}_{Tot} = -\frac{dE}{dx}_{Coll} + -\frac{dE}{dx}_{Brems}. \tag{4.1}$$

The derivations of energy loss formulas for these two processes are somewhat distinct. Although widely used, mean energy loss for travelling electrons is rarely measured directly but is obtained from energy loss or stopping power theories, hence the need for accurate derivation. Bethe's theory of energy loss (1930, 1932, 1934) [1], is the fundamental standard and 10 keV is commonly accepted as the lower limit for the

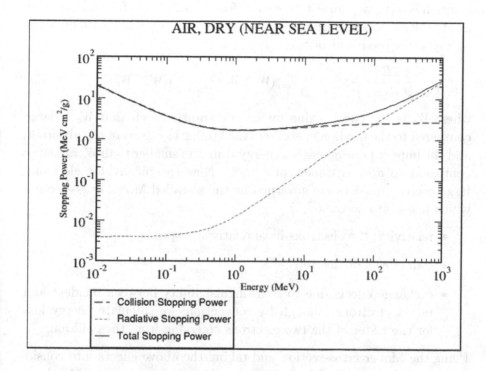

Figure 4.2 Energy Losses (ICRP data) for Electrons in Dry Air. The Collisional, Radiative and Total Stopping Power are plotted. The code used is from NIST, Online at https://dx.doi.org/10.18434/T4NC7P.

validity of the Bethe formula for electrons. We start out with the Bethe formula for energy loss and therefore the Bethe formula for stopping power with relativistic considerations but without the Bloch and other corrections (Barkas effect, density effect, etc.), given by Eq. 4.2. (in accord with ICRU Report 37 [8]):

$$\left| \frac{1}{\rho} \frac{dE}{dx} = \frac{1}{\rho} S_{col} = \frac{4\pi r_e^2 m_e c^2}{u} \frac{Z_1^2}{\beta^2} \frac{Z_2}{A_2} \times \left[\ln\left(\frac{2 m_e c^2 \beta^2}{I(1-\beta^2)} \right) - \beta^2 \right]. \right| \quad (4.2)$$

Eq. 4.2 is valid for all charged projectiles both light (e.g., electrons and positrons) and heavy (e.g., protons and alpha particles) as long as the requirements of first Born approximation are satisfied, within first order perturbation theory (see Appendix A.3). Moreover, Eq. 4.2 is a combined result following separate integration for distant (or soft) and near (or hard) collisions, with arbitrarily set cut-off points between them. That is, Eq. 4.2 is the result of two components:

$$\frac{1}{\rho}\frac{dE}{dx} = \frac{1}{\rho}S_{col} = \frac{1}{\rho}S_{col}(W < W_c) + \frac{1}{\rho}S_{col}(W > W_c) \qquad (4.3)$$

where W_c is the cutoff value for energy transfer such that W_c is large compared to the binding energies of the atomic electrons of the absorber; and the impact parameters for energy transfers smaller than W_c are large compared to atomic dimensions ([8]). Now specifically for electrons, large energy transfers are governed by the so-called Moller cross section which takes into account:

- relativity; the electrons have relativistic speeds;

- spin effects of electrons;

- exchange effects due to indistinguishability between incident and target electrons (though by convention, we calculate energy loss for the faster of the two electrons emerging from the collision.

Using the Moller cross-section and taking the above effects into consideration, and using the Bhadha cross section for positrons (which have no exchange effects as with electrons) the final electronic or collisional mass stopping power for electrons and positrons is given by [108]:

$$\boxed{\begin{aligned} \frac{1}{\rho}\frac{dE}{dx} &= \frac{1}{\rho}S_{col}^{\pm} \\ &= \frac{2\pi r_e^2 m_e c^2}{u}\frac{1}{\beta^2}\frac{Z_2}{A_2} \times \left[\ln\left(\frac{T}{I}\right)^2 + \ln(1+\tau/2) + F^{\pm}(\tau) - \delta\right], \end{aligned}}$$

$$(4.4)$$

where for electrons,

$$F^{-}(\tau) = (1-\beta^2)[1+\tau^2/8 - (2\tau+1)\ln 2] \qquad (4.5)$$

and for positrons,

$$F^{+}(\tau) = 2\ln 2 - \left(\frac{\beta^2}{12}\right) \times \left[23 + \frac{14}{(\tau+2)} + \frac{10}{(\tau+2)^2} + \frac{4}{(\tau+2)^3}\right] \qquad (4.6)$$

Other symbols in Eq. 4.4, Eq. 4.5 and Eq. 4.6 have the following meaning:

$$\tau = \frac{T}{(m_e c^2)} : \text{electron kinetic energy normalized by electron rest mass;} \quad (4.7)$$

$F^{\pm}$ stopping power function of electron and positron, and δ density effect parameter. For calculations in bulk, the electron density of the absorber, N_e given by:

$$N_e = \frac{N_A Z_2}{A_2}, \quad (4.8)$$

is used such that Eq. 4.4 becomes:

$$\boxed{\begin{aligned} \frac{1}{\rho}\frac{dE}{dx} &= \frac{1}{\rho}S_{col}^{\pm} \\ &= 2\pi r_e^2 m_e c^2 \frac{1}{\beta^2}\frac{N_A Z_2}{A_2} \times \left[\ln\left(\frac{T}{I}\right)^2 + \ln(1+\tau/2) + F^{\pm}(\tau) - \delta\right], \end{aligned}}$$

$$(4.9)$$

or in more compact form:

$$\boxed{\frac{1}{\rho}\frac{dE}{dx} = \frac{1}{\rho}S_{col}^{\pm} = C_e\frac{N_e}{\beta^2} \times \left[\ln\left(\frac{T}{I}\right)^2 + \ln(1+\tau/2) + F^{\pm}(\tau) - \delta\right],} \quad (4.10)$$

with

$$C_e = 2\pi r_e^2 m_e c^2 = 2\pi (2.8179 \times 10^{-13} \text{ cm})^2 (0.511 \text{ MeV})$$
$$= 2.54948 \times 10^{-25} \text{ MeV} \cdot \text{cm}^2 \quad (4.11)$$

Example 4.1 *What is the difference between the so-called restricted mass electronic (or collisional) stopping power $\frac{1}{\rho}S_{\Delta}^{\pm}$ and the mass electronic stopping power for electrons and positrons $\frac{1}{\rho}S_{col}^{\pm}$?*

Solution

Unlike the mass electronic stopping power for electrons and positrons $\frac{1}{\rho}S_{col}^{\pm}$ defined in Eq. 4.4, the restricted mass electronic (or collisional) stopping power $\frac{1}{\rho}S_\Delta^\pm$ is the mean energy loss by electrons and positrons per path length in a material due to ionization or excitation involving energy transfers up to a chosen cutoff energy Δ ([108]):

$$\frac{1}{\rho}S_\Delta^\pm = \frac{2\pi r_e^2 m_e c^2}{u}\frac{1}{\beta^2}\frac{Z_2}{A_2} \times \left[\ln\left(\frac{T}{I}\right)^2 + \ln(1+\tau/2) + H^\pm(\tau,\eta) - \delta\right], \quad (4.12)$$

where $\eta = \Delta/T$ is the fractional cutoff energy. For electrons:

$$H^-(\tau,\eta) = -1 - \beta^2 + \ln[4(1-\eta)\eta] + (1-\eta)^{-1}$$
$$+ (1-\beta^2)\left[\frac{\tau^2\eta^2}{2} + (2\tau+1)\ln(1-\eta)\right]. \quad (4.13)$$

For positrons,

$$H^+(\tau,\eta) = \ln(4\eta) - \beta^2\left[1 + (2-\xi^2)\eta - (3+\xi^2)\left(\frac{\xi\tau}{2}\right)\eta^2\right.$$
$$\left. + (1+\xi\tau)(\frac{\xi^2\tau^2}{3})\eta^3 - (\frac{\xi^3\tau^3}{4})\eta^4\right]. \quad (4.14)$$

Example 4.2 *From a theoretical mathematical physics viewpoint, show the utility of Eq. 4.10 in medical radiation dosimetry, the determination of absorbed dose to a medium, usually a patient's body in the final analysis.*

Solution

Eq. 4.10 is clearly needed in medical dosimetry due to the Bragg-Gray relationship (from Bragg-Gray cavity theory [74]) which is the basis for calculating the absorbed dose to an irradiated medium. In the case of a photon-irradiated *medium*, the dosimetry is done by measuring the ionisation caused by the photons in the gas volume of the dosimeter (usually an ion chamber) having a wall material different from the gas and different from the *medium* such that the absorbed dose is proportional to:

$$D_{med} \propto \frac{\left(\frac{dE}{\rho dx}\right)_{wall}}{\left(\frac{dE}{\rho dx}\right)_{gas}}\frac{\left(\frac{\mu_{en}}{\rho}\right)_{med}}{\left(\frac{\mu_{en}}{\rho}\right)_{wall}} + (1-f)\frac{\left(\frac{dE}{\rho dx}\right)_{med}}{\left(\frac{dE}{\rho dx}\right)_{gas}} \quad (4.15)$$

where $\frac{\mu_{en}}{\rho}$ is the photon mass-energy absorption coefficient for denoted material, f is the fraction of all ionisations in the chamber gas volume due to electrons arising from photon interactions within the chamber wall. The utility of the mass collisional stopping powers $\frac{dE}{\rho dx}$ of the wall, the gas, the medium (med) are all obvious in Eq. 4.15.

Example 4.3 *Calculate the collisional stopping power S_{col} for 1 MeV electrons incident on a lead absorber with $Z = 82$, $A_2 = 207.2\,g/mol$ and $I = 823\,eV$ using Eq. 4.10. Compare the result with data/graphs from NIST (https://physics.nist.gov/PhysRefData/Star/Text/ESTAR.html).*

Solution

We calculate step by step the various required variables in Eq. 4.10:

$$\frac{1}{\rho}\frac{dE}{dx} = \frac{1}{\rho}S_{col}^{\pm} = C_e \frac{N_e}{\beta^2} \times \left[\ln\left(\frac{T}{I}\right)^2 + \ln(1 + \tau/2) + F^{\pm}(\tau) - \delta \right].$$

From Eq. 4.11,

$$C_e = 2\pi r_e^2 m_e c^2 = 2\pi (2.8179 \times 10^{-13}\ \text{cm})^2 (0.511\ \text{MeV})$$
$$= 2.54948 \times 10^{-25}\ \text{MeV} \cdot \text{cm}^2$$

From Eq. 4.8,

$$N_e = \frac{N_A Z_2}{A} = \frac{6.022 \times 10^{23}\ \text{mol}^{-1} \times 82}{207.2\ \text{g mol}^{-1}} = 2.388 \times 10^{23}\ \text{electrons g}^{-1}$$

From Eq. 4.7

$$\tau = \frac{T}{(m_e c^2)} = \frac{1\ \text{Mev}}{0.511\ \text{MeV}} = 1.9569$$

From Eq. 3.30,

$$\beta^2 = \frac{v^2}{c^2} = 1 - \frac{1}{(1 + \frac{T}{m_e c^2})^2} = 1 - \frac{1}{(1 + \tau)^2} = 0.8856$$

$$1 - \beta^2 = 1 - 0.8856 = 0.1144$$

From Eq. 4.5

$$F^-(\tau) = (1 - \beta^2)[1 + \tau^2/8 - (2\tau + 1)\ln 2]$$

$$= (1 - 0.8856)[1 + 1.9569^2/8 - (2 \times 1.9569 + 1)\ln 2] = -0.22048.$$

Going to the NIST website (at `https://physics.nist.gov/PhysRefData/Star/Text/ESTAR.html`) and selecting lead, the density effect parameter δ for electrons at 1 MeV is 0.1809. Thus,

$$\frac{1}{\rho}\frac{dE}{dx} = \frac{1}{\rho}S_{col}^{\pm} = C_e\frac{N_e}{\beta^2} \times \left[\ln\left(\frac{T}{I}\right)^2 + \ln(1 + \tau/2) + F^{\pm}(\tau) - \delta\right],$$

$$= 2.54948 \times 10^{-25}\ \text{MeV}\cdot\text{cm}^2\frac{2.388 \times 10^{23}\ \text{electrons g}^{-1}}{0.8856}$$

$$\times \left[\ln\left(\frac{10^6}{823}\right)^2 + \ln(1 + 1.9569/2) + F^{-}(1.9569) - 0.1809\right],$$

$$= 0.994\ \text{MeV.cm}^2/\text{g}$$

Finally, the result using NIST data (at `https://physics.nist.gov/PhysRefData/Star/Text/ESTAR.html`) is 0.9939 MeV. cm^2/g which is the same as our result.

Exercise 53

Calculate the collisional stopping power S_{col} for electrons of the following kinetic energies incident on a lead absorber with $Z = 82$, $A_2 = 207.2$ g/mol and $I = 823$ eV using Eq. 4.10 and compare your results with data/graphs from NIST (`https://physics.nist.gov/PhysRefData/Star/Text/ESTAR.html`): (a) 5 MeV (b) 10 MeV (c) 15 MeV

Exercise 54

Particle physics experiments make use of water Cherenkov detectors where electrons are projected at energies above 0.8 MeV, with many applications. Water also closely resembles human tissue in terms of interactions with high energy electrons, with applications in radiation therapy. Calculate the collisional stopping power S_{col} for electrons of the following kinetic energies incident on liquid water as absorber with $Z_2 = 10$, $A_2 = 18.0153$ g/mol, (which gives electron density for water as:

$$n_e = \left(\frac{1\,\text{g}}{18.0153\,\text{g}}\right)(6.022 \times 10^{23}\,\frac{\text{molecules}}{\text{g}})(10\,\frac{\text{electrons}}{\text{molecule}})$$

$$= 3.343 \times 10^{23}\,\frac{\text{electrons}}{\text{g}},$$

and $I = 75$ eV using Eq. 4.10 and compare your results with data/graphs from NIST (`https://physics.nist.gov/PhysRefData/Star/Text/ESTAR.html`):// (a) 1 MeV (b) 5 MeV (c) 10 MeV (d) 15 MeV

4.2 RADIATIVE LOSS VIA BREMSSTRAHLUNG FOR LIGHT CHARGED PARTICLES

As already adumbrated, the Bethe-Heitler formula for energy loss via bremsstrahlung is valid for light and heavy charged particles as long as the conditions of derivation such as Born's approximation and first order perturbation, are satisfied. When these are not satisfied, other approximations such as the Thomas-Fermi approximation are used. A Hartree-Folk approach is another alternative. Although it is easier to measure, there is no simple analytic expression for calculating energy loss via bremsstrahlung. We present the approximate formula here and apply some restrictions and corrections for electrons and positrons:

$$-\frac{dE}{dx}\bigg|_{Brems} \approx 4\alpha N_A \frac{e^4}{m_1^2 c^4}\left(\ln\frac{183}{Z_2^{1/3}}\right)\frac{Z_2(Z_2+1)}{A_2}q_1^2 E. \qquad \text{(3.80 revisited)}$$

Again, in Eq. 3.80, the subscript 2 refers to the properties of the material and the subscript 1 refers to the properties of the incoming charged projectile, α is the fine structure constant, q is electric charge in units of e while E is the energy of the incoming projectile. Here, we focus on electrons and positrons as the incoming projectiles. Note that the energy loss is about the same for both electrons and positrons. This is because the energy transfer is about the same whether the interaction involves a repulsive (electron-orbital electron) or attractive (positron-orbital electron) force. No wonder that the tracks of positrons in a material are similar to those of electrons and the ranges are also roughly the same, for the same initial energies. However, for positrons, towards the end of the range, a positron combines with a negative electron in the material and the two particles annihilate, with the creation of two photons escaping in opposite directions. To analyse the energy loss for electrons, positrons, muons and other light particles, it is usual to introduce a so-called *radiation length X_0*:

$$\frac{1}{X_0} = 4\alpha N_A \frac{e^4}{m_e^2 c^4}\left(\ln\frac{183}{Z_2^{1/3}}\right)\frac{Z_2(Z_2+1)}{A_2}. \qquad (4.16)$$

Note that the mass of the incoming projectile m_1 has been replaced with the mass of the incoming electron m_e in X_0. Thus for electrons,

$$-\frac{dE}{dx}\bigg|_{Brems} \approx \frac{1}{X_0}E. \qquad (4.17)$$

The radiation length only depends on the electron and some material constants. The radiation length characterises the radiation of electrons in the target material. Thus, tables of radiation lengths for various materials and different energies are available in the literature and online repositories such as those of the Particle Data Group (see PDG) It is interesting to note the exponential function apparent in Eq. 4.17:

$$-\frac{dE}{dx}_{Brems} \approx \frac{1}{X_0}E \implies E(x) = E_0 e^{-\frac{x}{X_0}}, \tag{4.18}$$

where x is the thickness of the absorber or target material. Note that X_0 has length dimension in Eq. 4.18, but can have dimensions of mass/area when obtained from mass radiative stopping power:

$$-\frac{dE}{\rho dx}_{Brems} \approx \frac{1}{X_0}E \implies E(x) = E_0 e^{-\frac{x}{X_0}}. \tag{4.19}$$

Due to bremsstrahlung, the energy of the projectile decreases exponentially as a function of x. Of course, after $x = X_0$, E_{X_0} becomes E_0/e or $0.37E_0$. These are very handy realizations for experiments with and applications of bremsstrahlung. Along these lines, another important parameter is the critical energy E_c which is the energy at which energy losses due to ionization (mainly collisional) and bremsstrahlung are equal. Thus:

$$-\frac{dE}{dx}_{ion}(E_c) = -\frac{dE}{dx}_{Brems}(E_c), \tag{4.20}$$

Tables of radiation lengths and critical energies are available for various elements and substances, illustrating the utilities of the formulas of chapter 3 and this chapter 4. Fig. 4.3 shows both the collisional (ionization) and radiative energy losses (stopping power) of the eye lens based on tables from NIST. The critical energy for water is about 80 MeV (see Particle Data Group, PDG, Atomic and Nuclear Properties of Materials) and it is not surprising that the eye lens has the radiative and collisional stopping powers equal to about 100 MeV (Fig. 4.3). Plots such as Fig. 4.3 which reveal the critical energies for various charged particles in different substances are used for the identification of the particles themselves. Furthermore, Fig. 4.3 and the dependencies shown in the formulas so far, illustrate that above tens of MeV, radiative loss (bremsstrahlung) is the dominant process for electrons. This is also true of positrons. For muons and pions, this dominance happens around a

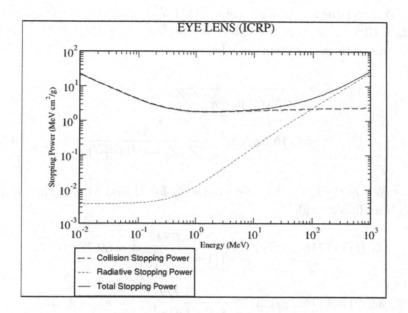

Figure 4.3 Stopping Power of Eye Lens (ICRP data) for Electrons. The Collisional, Radiative and Total Stopping Power are plotted. The code used is from NIST, NIST, available online.

few 100 GeV. At these very high energies, the dominance of radiative over collisional energy losses gives rise to electron-photon cascade showers. These showers arise from the high energy photons emitted. These in turn produce Compton electrons and electron-positron pairs which produce additional bremsstrahlung photons and the cascade continues into showers.

Example 4.4 *Calculate the radiation length X_0 for electrons in water.*

Solution

Eq. 4.16 can be recast more directly to give the radiation length:

$$X_0 = \frac{1}{4\alpha N_A \frac{e^4}{m_e^2 c^4} \left(\ln \frac{183}{Z_2^{1/3}} \right) \frac{Z_2(Z_2+1)}{A_2}} = \frac{1}{4\alpha N_A r_e^2 \left(\ln \frac{183}{Z_2^{1/3}} \right) \frac{Z_2(Z_2+1)}{A_2}} \quad (4.21)$$

where α is the fine structure constant $(1/137)$, N_A is Avogadro's constant (6.022×10^{23}), and r_e is the classical electron radius (2.818 fm). Using the values of these constants, we have:

$$X_0 = \frac{1}{4\alpha N_A r_e^2 \left(\ln \frac{183}{Z_2^{1/3}} \right) \frac{Z_2(Z_2+1)}{A_2}}$$

$$= (716.2 \text{ g/cm}^2) \times \frac{A_2}{Z_2(Z_2+1)\left(\ln \frac{183}{Z_2^{1/3}} \right)} \qquad (4.22)$$

To get X_0 for H_2O, we first determine X_0 for H and O and then apply Bragg's additivity rule.

$$X_0(H) = (716.2 \text{ g/cm}^2) \times \frac{1.00784}{1(1+1)\left(\ln \frac{183}{1^{1/3}} \right)} = 69.28 \text{ g/cm}^2.$$

$$X_0(O) = (716.2 \text{ g/cm}^2) \times \frac{15.999}{8(8+1)\left(\ln \frac{183}{8^{1/3}} \right)} = 35.24 \text{ g/cm}^2.$$

Bragg's additivity rule applied to X_0 is:

$$\frac{1}{X_0} = \sum_i \frac{w_i}{X_i} \qquad (4.23)$$

where w_i is the weight fraction and X_i the radiation length for the ith element or component. Thus,

$$\frac{1}{X_0(H_2O)} = \frac{2}{18} \times \frac{1}{X_0(H)} + \frac{16}{18} \times \frac{1}{X_0(O)}$$

$$= \left(\frac{2}{18 \times 69.28} \right) + \left(\frac{16}{18 \times 35.24} \right) = 0.02525 \text{ cm}^2/\text{g}$$

Finally,

$$X_0(H_2O) = 39.6 \text{ g/cm}^2.$$

Exercise 55

Consider air as a mixture that is 75.5% nitrogen $(Z_2 = 7)$, 23.2% oxygen $(Z_2 = 8)$, and 1.3% argon $Z_2 = 18$ and calculate the radiation length for electrons in air using Eqs. 4.22 and 4.23.

Exercise 56

Obviously, the dimension of the radiation length X_0 in Eq. 4.18 is length and in Eq. 4.16 it is mass/area due to the presence of density ρ. What is the radiation length for electrons in water X_0 in cm, using the result of Example 4.4.

Exercise 57

What is the radiation length for electrons in air X_0 in cm, using the result of Exercise 55.

Exercise 58

The right hand side of Eq. 4.18 or Eq. 4.19 can be derived explicitly thus:

$$-\frac{dE}{dx}_{Brems} \approx \frac{1}{X_0}E \implies \frac{dE}{E} = -\frac{dx}{X_0}. \tag{4.24}$$

Hence,

$$\int_{E_0}^{E} \frac{dE}{E} = \frac{-1}{X_0} \int_0^x dx \implies E(x) = E_0 e^{-\frac{x}{X_0}}. \tag{4.25}$$

Find the values of $\frac{E(x)}{E_0}$ for

(a) $x = (1/2)X_0$
(b) $x = X_0$
(c) $x = 2X_0$

The above derivations apply most accurately to monoenergetic beams of electrons, positrons, muons and other light charged particles. However, in the case of beta particles from a radioactive nucleus following beta decay, we state the following relations with respect to energy loss via bremsstrahlung creation. The potential for bremsstrahlung is directly proportional to projectile energy as well as the atomic number Z of the absorber. For beta particles, the fraction of the beta particle energy Q_β converted to X-ray photons is:

$$f_\beta = 3.5 \times 10^{-4} ZQ_\beta. \tag{4.26}$$

Note that Q_β in Eq. 4.26 is the maximum kinetic energy or endpoint of the beta particle $Q_{\beta max}$. But the fraction of electron energy from a monoenergetic electron source, such as a linear accelerator, which gets converted to bremsstrahlung photons is

$$f_e = 10^{-4} ZE_e, \tag{4.27}$$

where E_e is the electron energy in MeV.

Just as collisional energy loss leads to collisional stopping powers for materials, radiative energy losses engender radiative stopping powers of materials. The mass radiative stopping powers for electrons and positrons can be written as [8, 108]:

$$\frac{dE}{\rho dx}\bigg|_{Brems} = \frac{1}{\rho}S_{rad} = \alpha r_e^2 \frac{N_A}{uA_2}(E+m_ec^2)[Z_2^2\phi_{rad,n}+Z_2\phi_{rad,e}], \quad (4.28)$$

where E is the kinetic energy of the electron or positron, α is the fine structure constant, and ϕ_{rad} denotes the scaled, dimensionless radiative energy loss cross-sections in two components as follows:

$$\phi_{rad,n} = \frac{(\alpha r_e^2 Z_2^2)^{-1}}{(E+m_ec^2)}\int k\frac{d\sigma_n}{dk}dk, \quad (4.29)$$

for electron-nuclear bremsstrahlung, with $\frac{d\sigma_n}{dk}$ denoting the bremsstrahlung emission cross section in the screened Coulomb field of the atomic nucleus, for emitted photon energy k; and

$$\phi_{rad,e} = \frac{(\alpha r_e^2 Z_2^2)^{-1}}{(E+m_ec^2)}\int k\frac{d\sigma_e}{dk}dk, \quad (4.30)$$

for electron-electron bremsstrahlung, where $\frac{d\sigma_e}{dk}$ denotes the bremsstrahlung emission cross section in the screened Coulomb field of the atomic nucleus, for emitted photon energy k. The cross-sections are energy-dependent [7, 8, 109] and data tables are available [108]. Owing to the lack of more solid theoretical framework, the mass radiative stopping power for compounds and mixtures, is usually obtained using Bragg's additivity rule:

$$\frac{dE}{\rho dx}\bigg|_{Brems} = \frac{1}{\rho}S_{rad} = \sum_i w_i[\frac{1}{\rho}S_{rad}]_i \quad (4.31)$$

where w_i denotes weight fraction and $[\frac{1}{\rho}S_{rad}]_i$ is the radiative mass stopping power of the ith atomic constituent.

Just as the total energy loss for light charged particles is equal to the sum of both the collisional and radiative losses (Eq. 4.1), the total mass stopping power of a material for light charged particles is the sum of both:

$$\frac{dE}{\rho dx}\bigg|_{Tot} = \frac{1}{\rho}S_{Tot} = \frac{dE}{\rho dx}\bigg|_{Coll} + \frac{dE}{\rho dx}\bigg|_{Brems} = \frac{1}{\rho}S_{Coll} + \frac{1}{\rho}S_{Rad} \quad (4.32)$$

Eq. 4.32 implies that

$$S_{Coll}(E_c) = S_{Rad}(E_c) = \frac{1}{2}S_{Tot}. \tag{4.33}$$

Example 4.5 *We defined the critical energy E_c in Eq. 4.20 as the kinetic energy of the light charged particle at which S_{rad} and S_{coll} are equal. An empirical relation for estimating E_c for a given absorber material Z_2 is:*

$$E_c = \frac{constant}{Z_2} = \frac{800 \ MeV}{Z_2}. \tag{4.34}$$

Check the veracity of Eq. 4.34 for Carbon, Copper and Lead using figures from NIST (https://physics.nist.gov/PhysRefData/Star/Text/ESTAR. html) such as Fig. 4.3 and Fig. 4.2 and comment on the trend.

Solution

For Carbon with $Z_2 = 6$, then Eq. 4.34 gives:

$$E_c = \frac{800 \ MeV}{Z_2} = \frac{800 \ MeV}{6} = 133 \ MeV.$$

From the NIST data/graph, $E_c = 96$ MeV ≈ 100MeV For Copper with $Z_2 = 29$, then Eq. 4.34 gives:

$$E_c = \frac{800 \ MeV}{Z_2} = \frac{800 \ MeV}{29} = 27.6 \ MeV.$$

From the NIST data/graph, $E_c = 24.3$ MeV ≈ 24 MeV
   For Lead with $Z_2 = 82$, then Eq. 4.34 gives:

$$E_c = \frac{800 \ MeV}{Z_2} = \frac{800 \ MeV}{82} = 9.8 \ MeV.$$

From the NIST data/graph, $E_c = 10$ MeV. Thus, the trend is that of increasing agreement between the NIST data/graph as Z_2 increases.

Example 4.6 *Another useful empirical relation involving the critical energy E_c is the ratio of collisional and radiative stopping powers:*

$$\frac{S_{coll}}{S_{rad}} = \frac{E_c}{E} \tag{4.35}$$

where E is the kinetic energy of the light charged particle. Check the veracity of Eq. 4.35 for Copper at 1 MeV, 10 MeV and 100 MeV.

Solution

Using data from NIST (https://physics.nist.gov/PhysRefData/Star/Text/ESTAR.html): for 1 MeV electron incident on Cu $Z_2 = 29$,

$$\frac{S_{coll}}{S_{rad}} = \frac{1.236}{0.0458} = 27$$

$$\frac{E_c}{E} = \frac{24.3}{1} = 24.3;$$

for 10 MeV electron incident on Cu,

$$\frac{S_{coll}}{S_{rad}} = \frac{1.436}{0.565} = 2.54$$

$$\frac{E_c}{E} = \frac{24.3}{10} = 2.43;$$

for 100 MeV electron incident on Cu,

$$\frac{S_{coll}}{S_{rad}} = \frac{1.661}{7.079} = 0.235$$

$$\frac{E_c}{E} = \frac{24.3}{100} = 0.243.$$

The trend is obvious; the approximation (Eq. 4.35) improves as the energy increases.

Exercise 59 .

Find the critical energy using Eq. 4.34 for Silver $(Z_2 = 47)$ and Uranium $(Z_2 = 92)$ and compare with results using figures from NIST (https://physics.nist.gov/PhysRefData/Star/Text/ESTAR.html).

Exercise 60

Check the veracity of Eq. 4.35 for Carbon at 1 MeV, 10 MeV and 100 MeV.

4.3 RANGE OF LIGHT CHARGED PARTICLES

Light charged projectiles such as electrons do not have straight paths in target materials due to scatterings from atomic electrons and nuclei. Rather, they travel in zig-zag paths which put together are longer than the simple range we had for heavy charged particles. In other words, the *distance* between the entrance point and final (absorption) point of an incident particle in the absorber or target medium is longer. Hence, an empirical quantity called *continuous slowing down approximation* (CSDA) [8, 109] is used. For electrons and positrons, the CSDA range R_{CSDA} is given as:

$$R_{CSDA}\left[\frac{g}{cm^2}\right] = \int_0^{E_0} \left(\frac{dE}{\rho dx}\right)^{-1} dE, \qquad (4.36)$$

where E_0 is the initial kinetic energy of the electron. Since $\left(\frac{dE}{\rho dx}\right)$ has units $MeV \cdot cm^2 \cdot g^{-1}$, R_{CSDA} has units $g \cdot cm^{-2}$. The approximation consists in assuming that the rate of energy loss at every point along the track is equal to the total stopping power. That is, fluctuations in energy-loss are neglected. Simple empirical relations have been developed for calculation of CSDA range and electron energy loss [114]. Monte Carlo methods are also used and are in fact increasingly preferred in calculating the paths of electrons and positrons in materials [85] owing to increasing capabilities in machine computation. The maximum range R_{max} of beta particles (electrons and positrons) can be estimated and expressed in a material-independent manner, with units of $\frac{g}{cm^2}$. Katz and Penfold [58] worked an empirical formula that has led to a useful rule of thumb. They presented R_{max} with respect to the end point beta energy Q_β (discussed in chapter one) thus:

$$R_{max}\left[\frac{g}{cm^2}\right] = \begin{cases} 0.412Q_\beta^{1.265-0.0954\ln Q_\beta}, & \text{if } 0.01 \leq Q_\beta \leq 2.5\,\text{MeV}. \\ 0.530Q_\beta - 0.016, & \text{if } Q_\beta > 2.5\,\text{MeV}. \end{cases} \qquad (4.37)$$

The above empirical equations corroborate the useful rule of thumb common in medical physics: the range of a beta particle in $\frac{g}{cm^2}$ is approximately 0.5 of its endpoint energy in MeV. Of course, the absorber's ability to stop the beta particles depends on the number of electrons per area of the absorber (electron areal density). Hence, the range or density thickness $\left(\frac{g}{cm^2}\right)$ of a material becomes a generic quantifier by which various absorbers can be compared. Once the maximum range is known, practical thicknesses t in unit of length, such as thickness of shielding, penetration depth for radiotherapy, etc, can be calculated thus:

$$t[cm] = \frac{R_{max}}{\rho}, \qquad (4.38)$$

where ρ is the density of the material.

Example 4.7 *What is the thickness of Copper in cm, needed to completely stop beta particles produced via the decay of Co-60, given that the maximum decay energy of these beta particles from Co-60 is 0.3179 MeV?*

Solution

We need to calculate R_{max}, so using Eq. 4.37, we have,

$$(0.412)(0.3179)^{1.265-0.0954\ln 0.3179} = 0.08529 \frac{g}{cm^2}.$$

Finally, we insert the density of Copper, $\rho = 8.933 \frac{g}{cm^3}$ into Eq. 4.38, to get the required thickness:

$$t[cm] = \frac{R_{max}}{\rho} = \frac{0.08529 \frac{g}{cm^2}}{8.933 \frac{g}{cm^3}} = 0.00955 \, cm.$$

Example 4.8 *What is the range in air of beta particles with a maximum energy of 3.8 MeV? Take density of air as 0.0013 g/cc (where cc is cubic centimeters).*

Solution

We need to calculate $t[cm]$. We can estimate the density thickness R_{max} using Eq. 4.37 for energy greater than 2.5 MeV:

$$(0.53 \times 3.8 \, MeV) - 0.016 = 1.998 \frac{g}{cm^2}.$$

Now,

$$t[cm] = \frac{R_{max}}{\rho} = \frac{1.998 \frac{g}{cm^2}}{0.0013 \frac{g}{cm^3}} = 1537 \, cm$$

Exercise 61

The range of a beta particle in $\frac{g}{cm^2}$ is approximately 0.5 of its endpoint energy in MeV. What is the approximate range in cm for beta particles of Sr-90 with an endpoint energy of 2.2 MeV when incident on tissue with a density of 1.0599 g/cm^3?

4.3.1   Radiation Yield

The fraction of the initial kinetic energy, E_0 of the light charged particle that is converted to bremsstrahlung energy as the charged particle slows down to rest, is defined as the *radiation yield*, $Y(E_0)$ and in the CSDA range, it is given by [108]:

$$Y(E_0) = \frac{1}{E_0} \int_0^{E_0} \frac{S_{rad}(E)}{S_{Tot}(E)} dE. \qquad (4.39)$$

Note that the radiation yield calculated with Eq. 4.39 may be less than it actually is due to the inability of Eq. 4.39 to account for electron straggling, which contemporary Monte Carlo codes, for instance, do account for [100]. Interestingly, on average, the mass collisional stopping power of an absorber, $\frac{1}{\rho}S_{Coll}$ can be related with the radiation yield, $Y(E_0)$, the CSDA range R_{CSDA} and the kinetic energy of light charged particles based on the respective definitions for these quantities, thus:

$$\left(\frac{1}{\rho}S_{Coll}\right)_{avg} = \frac{(1 - Y(E_0))}{R_{CSDA}(E_0)} E_0. \qquad (4.40)$$

Exercise 62

Estimate the $Y(E_0)$ for a beam of electrons incident on liquid water at 10 MeV. Compare the result with the NIST data (at https://physics.nist.gov/PhysRefData/Star/Text/ESTAR.html) which is 0.04072. That is, only 4% of 10 MeV or 0.4072 MeV is radiated from a 10 MeV electron beam striking a water absorber. Hint: The ratio $\frac{S_{rad}(E)}{S_{Tot}(E)}$ in Eq. 4.39 is what is needed but it is hardly available analytically, hence, use numerical summation for the estimation:

$$Y(E_0) = \frac{1}{E_0} \int_0^{E_0} \frac{S_{rad}(E)}{S_{Tot}(E)} dE \approx \frac{1}{E_0} \sum_{i=1}^{n} \left(\frac{S_{rad}(E)}{S_{Tot}(E)}\right)_{i,avg} \Delta E. \qquad (4.41)$$

where $\left(\frac{S_{rad}(E)}{S_{Tot}(E)}\right)_{i,avg}$ is the average of $\frac{S_{rad}(E)}{S_{Tot}(E)}_i$ for the ith interval and $\left(\frac{S_{rad}(E)}{S_{Tot}(E)}\right)_{i,avg} \Delta E$ is the area of the ith interval.

4.4   ESTAR AND OTHER SOFTWARE PACKAGES

As stated in Chapter 3, the National Institute of Standards and Technology (NIST) website provides tables and formulas for calculation of ranges and stopping powers for electrons (ESTAR), protons (PSTAR) and alpha particles (ASTAR), see https://physics.nist. gov/PhysRefData/Star/Text/intro.html. As noted already, ESTAR: Electronic Stopping Power and Range; PSTAR: Proton Stopping Power and Range; and ASTAR: Alpha Particle Stopping Power and Range all provide data for the calculation of mass stopping power and ranges of charged particles in matter.

The ESTAR (Electronic Stopping Power and Range) program calculates the stopping power, density effect parameters, the CSDA range, and radiation yield tables for electrons in 98 elements from the Periodic Table and various materials such as blood, bone, glass, tissue, water and wax.

Another software package is SRIM (Stopping and Range of Ions in Matter), a software package that has thousands of users around the world. It has been continuously upgraded since its introduction in 1985 by James Ziegler. A recent textbook, "SRIM-The Stopping and Range of Ions in Matter" describes in detail the fundamental physics of the software. [129]". Details can be obtained from SRIM (with url:http://www.srim.org). There is also TRIM (the Transport of Ions in Matter) which is closely connected with SRIM. TRIM calculates 3D distributions of ions in up to 8 layers of selected materials as well as certain kinetic phenomena associated with the ion's energy loss such as ionization and phonon production. Furthermore, there are codes for evaluating the various corrections to the Bethe formula: Bloch, Barkas and Shell corrections, such as the BEST (BEthe STopping) developed by M. J. Berger and H. Bichsel [16] and used extensively in ICRU 49 [7]. The BEST includes the evaluation of the density-effect correction.

Example 4.9 *Use the ESTAR (Electronic Stopping Power and Range) program to estimate the CSDA range of 0.3179 MeV electrons in Copper and make a cursory comparison with the result in Example 4.7. Note the differences in formalism between the calculations.*

Solution

Figure 4.4 shows the output from ESTAR for electrons travelling in Copper. The kinetic energy column shows data that include the 0.3179

(required) Kinetic Energy (MeV)	Stopping Power (MeV cm²/g)			CSDA Range (g/cm²)	Radiation Yield	Density Effect Parameter
	Collision	Radiative	Total			
2.500E-01	1.691E+00	2.045E-02	1.711E+00	9.603E-02	6.981E-03	1.236E-01
3.000E-01	1.579E+00	2.172E-02	1.601E+00	1.263E-01	7.945E-03	1.604E-01
3.500E-01	1.501E+00	2.307E-02	1.524E+00	1.584E-01	8.860E-03	1.958E-01
4.000E-01	1.444E+00	2.450E-02	1.469E+00	1.918E-01	9.741E-03	2.302E-01

Figure 4.4 CSDA Range of Electrons in Copper using NIST's ESTAR Code, Online.

MeV of the beta particles. Extrapolation may be needed for more exact comparison. Both results are within the same order of magnitude and are about 0.01 cm. However, even a cursory comparison shows that the CSDA Range of electrons from a mono-energetic beam of energy equal to a given Q_β, here, 0.3179 MeV, is greater than that of the beta particle. The reason is simple: beta particles have a continuous energy spectrum from zero energy up to a maximum kinetic energy as explained and illustrated in Chapter 1. Another relevant rule of thumb is that the average beta decay energy $\bar{Q}_\beta$ is about one-third of the maximum beta decay energy Q_β and $\bar{Q}_\beta$ is sometimes used in calculations to have comparable results with mono-energetic beams of electrons.

4.5 MULTIPLE COULOMB SCATTERING AND GAUSSIAN APPROXIMATIONS

Charged projectiles traversing matter undergo not only inelastic collisions with atomic electrons of the absorber but also undergo multiple elastic scattering or multiple Coulomb scattering from nuclei of absorber whereby only the directions of the projectiles change. This leads to the total path length travelled becoming longer than the penetration depth and this elongation can be significant for electrons [9, 119, 120]. When the thickness of the absorber is large enough, the number of multiple scattering events becomes significant (usually > 20) such that the angular dispersion can be modelled as a Gaussian, at small angles. Such large absorbers are practically important as in the case of tissue in radiotherapy. In the physics of radiotherapy, attempts are being made to use multiple Coulomb scattering for dose calculations [56]. For mathematical buildup, single scattering is given by the Rutherford formula [102] for elastic scattering from nuclei:

$$\frac{d\sigma}{d\Omega} = \frac{r_e^2 Z_1^2 Z_2^2}{4} \left(\frac{m_e c}{\beta p}\right)^2 \frac{1}{\sin^4 \theta/2} \tag{4.42}$$

There are theories that account for a wide range of the experimental distributions obtained through multiple scattering such as Moliere theory [15,78], Lewis theory and Lynch theory. For the Gaussian approximation, the angular distribution in the spatial domain has a central part that can be approximated by:

$$P(\theta)d\Omega \approx \frac{1}{2\pi\theta_0^2}e^{-\frac{\theta^2}{2\theta_0^2}}d\Omega \qquad (4.43)$$

where

$$\theta_0 = \frac{13.6\,MeV}{\beta pc}z\sqrt{\frac{l}{X_0}}\left[1+0.038\ln\frac{l}{X_0}\right]$$

and $\frac{l}{X_0}$ is the thickness of the absorber or medium measured in radiation lengths X_0. This expression for θ is obtained from fits to the Moliere distribution [78] and there are similar ones from other theories of multiple scattering mentioned above.

4.6   TAMM-FRANK-CERENKOV RADIATION FORMULA

Pavel Cerenkov (also spelt Cherenkov) in 1934 discovered that a charged particle moving in a dielectric medium with a velocity $v > c/n$, where c/n is the speed of light (specifically phase velocity) in the medium, emits a characteristic bluish-white radiation (UV plus visible light) which is now known as **Cerenkov radiation**. Ilya Frank and Igor Tamm gave the theoretical explanation in 1937. We focus here on the mathematics involved in the theoretical explanation. In 1958, Cerenkov, Frank and Tamm were awarded the Nobel Prize for discovery and explanation of the effect. Some features of Cerenkov radiation have applications such as particle identification in experimental high energy physics, lumines-cence imaging of tissues during radiotherapy [27, 75], ring imaging in Cerenkov counters, etc. Cerenkov radiation is emitted if the velocity of the projectile v is greater than the velocity of light in the medium:

$$v > \frac{c}{n}, \qquad (4.44)$$

where c is the speed of light in vacuum and n is the refractive index of the material. This leads to an electromagnetic version of the shock waves created whenever an object travels through air at a speed faster than the speed of sound in air. Here, the source of the sound wave (travelling object) or the source of the electromagnetic radiation (charged projectile)

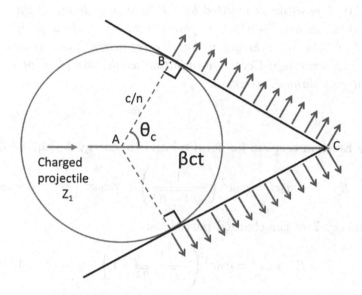

Figure 4.5 Schematic of the geometry of Cerenkov radiation for the non-dispersive (ideal) case.

is moving faster than the sound/light waves it creates, thereby leading the advancing wavefront. Just like a sonic boom, the electromagnetic shock wave develops with a coherent wave front that has a conical shape and the photons are emitted within an angle called the Cerenkov angle θ_c as shown in Fig. 4.5 which leads to the Cerenkov relation given by:

$$\cos\theta_c = \frac{AB}{AC} = \frac{\frac{c}{n}t}{\beta ct} = \frac{1}{\beta n}, \tag{4.45}$$

where $\beta = \frac{v}{c}$.

Thus, as the charged projectile slows down in the medium, θ_c is reduced. Since β and $\cos\theta$ are between 0 and 1, the following mathematical consequences ensue for the Cerenkov relation:

$$\theta_{c_{min}} = \cos^{-1}\left(\frac{1}{\beta n}\right) = \cos^{-1} 1 = 0. \tag{4.46}$$

$$\theta_{c_{max}} = \lim_{\beta\to 1}\theta_c = \lim_{\beta\to 1}\cos^{-1}\left(\frac{1}{\beta n}\right) = \cos^{-1}\frac{1}{n}. \tag{4.47}$$

Example 4.10 *A positron is emitted by* ^{18}F *beta-plus decay. If the kinetic energy of the positron is 635 keV, find (a) the speed of the positron. (b) The speed of light in water given that the refractive index of water* n_w *is 1.33. (c) Determine if Cerenkov radiation would take place or not, if the positron goes through water.*

Solution

(a) Using the Einstein relation for total relativistic energy E, Eq. 4.48,

$$E = T + m_0c^2 = m_0c^2 \left(\frac{1}{\sqrt{1-\beta^2}} \right) = \gamma m_0 c^2 \qquad (4.48)$$

the Kinetic energy T of the charged particle is:

$$T = E - m_0c^2 = m_0c^2 \left(\frac{1}{\sqrt{1-\beta^2}} - 1 \right) \qquad (4.49)$$

and substituting T with 635 keV, we obtain the speed of the positron as $v = 0.895c$.

(b) The speed of light in water is obtained thus:

$$v = \frac{c}{n} = \frac{c}{1.33} = 0.752c.$$

(c) Since the speed of the positron in water ($0.895c$) is greater than the speed of light in water ($0.752c$), Cerenkov radiation will take place.

Example 4.11 *(a) In terms of the refractive index of a medium, what is the threshold energy in MeV for the production of Cherenkov radiation by a high-speed (relativistic) electron traversing that medium? (b) Why are solid state Cherenkov radiation detectors made from high-index of refraction materials such as glass?*

Solution

Using Eq. 4.49, and replacing v with $\frac{c}{n}$, since $\beta = \frac{1}{n}$ and $\beta^2 = \frac{1}{n^2}$,

$$T = m_0c^2 \left(\frac{1}{\sqrt{1-\frac{1}{n^2}}} - 1 \right) = 0.511 \left(\frac{1}{\sqrt{1-\frac{1}{n^2}}} - 1 \right) \text{ MeV.} \qquad (4.50)$$

For water, this is about 0.264 MeV. (b) From Eq. 4.50, we see that the emission of Cherenkov radiation is favoured by high-index of refraction materials. These are very useful in high energy physics experiments.

Exercise 63

Calculate the electron threshold energy for a Cerenkov counter whose index of refraction is 1.6.

4.7 TRANSITION RADIATION

As a charged particle crosses the boundary of two media with different dielectric constants, it emits electromagnetic radiation, called transition radiation. It was discovered and explained in 1946 by Ginsburg and Frank. A good theoretical description can be found in [53]. The main reason is that in the material with low dielectric constant ε_1, the polarization effect is small so the electric field of the projectile, E_1 has a larger extent than in the material with higher dielectric constant, say, ε_2 where the polarization effect is greater. Thus, in this case, $E_1 > E_2$. The adaptation or reallocation of charges and electric fields lead to the transition radiation. A charged projectile with charge z in units of e, moving from vacuum into a material with plasma frequency ω_p emits transition radiation with energy E_t given by:

$$E_t = \frac{1}{3}\alpha z^2 \gamma \hbar \omega_p.\qquad (4.51)$$

Measurements of E_t enable the determination of γ which then yields the velocity of the particle. Andronic and Wessels [2] discuss the recent use of transition radiation for particle identification. The dependence of transition radiation on γ seen in Eq. 4.51 gives away the fact that there is a wide momentum range (1-100 GeV/c) where electrons and positrons are the only charged particles producing transition radiation. Thus, we consider it a form of energy loss more important for light charged particles than heavy charged particles.

Exercise 64

High-energy linear accelerators (linacs) used in radiotherapy often provide several electron beam energies ranging from 4 MeV to 25 MeV in addition to photon energies. These electron beams interact with the photon collimators, electron cones (applicators) and the patient, leading to the production of bremsstrahlung radiation. In addition, the electrons interact with the atoms of the absorbing materials (patient, collimator, etc.) via Coulomb force interactions including: inelastic collisions with orbital electrons of the target atoms; inelastic collisions with nuclei of the target atoms; elastic collisions with orbital electrons of the target atoms;

elastic collisions with nuclei of the target. The incident electrons lose their kinetic energy via ionization collisions and radiative collisions owing to the above processes. Find using online graphs linked in this chapter the average rate of energy loss (in MeV/cm) for a therapy electron beam in water and water-like tissues.

Exercise 65

A 12 MeV beam of electrons is incident on a body of water. At what depth will the energy be about 7 MeV. Hint: average rate of energy loss in water for electrons is 2 MeV/cm.

4.8 ANSWERS

Answer of exercise 53

Ans: (a) $1.12 \, \text{MeV.cm}^2/\text{g}$, (b) $1.201 \, \text{MeV.cm}^2/\text{g}$ and (c) $1.246 \, \text{MeV.cm}^2/\text{g}$.

Answer of exercise 54

Ans: (a) $1.849 \, \text{MeV.cm}^2/\text{g}$, (b) $1.892 \, \text{MeV.cm}^2/\text{g}$, (c) $1.968 \, \text{MeV.cm}^2/\text{g}$ and (d) $2.014 \, \text{MeV.cm}^2/\text{g}$.

Answer of exercise 55

Ans: $X_0(air) = 37.8 \, \text{g/cm}^2$ at STP.

Answer of exercise 56

Ans: 39.6 cm.

Answer of exercise 57

Ans: 29234 cm at STP.

Answer of exercise 58

Ans: (a) 0.607, (b) 0.368, (c) 0.135.

Answer of exercise 59

Ans: Silver: calculation 17 MeV and NIST data 16 MeV. Uranium: calculation 8.7 MeV and NIST data 9 MeV.

Answer of exercise 60

Ans: 1 MeV, Scol/Srad = 160 and Ec/E = 96, at 10 MeV, Scol/Srad

= 12, Ec/E = 9.6, and at 100 MeV Scol/Srad = 1 and Ec/E = 0.96. Same trend; greater correspondence at higher energy.

Answer of exercise 61

Ans: 1 cm

Answer of exercise 62

Ans: See NIST result given already.

Answer of exercise 63

Ans: 0.143 MeV.

Answer of exercise 64

Ans: 2 MeV/cm. This is a good rule of thumb in medical physics.

Answer of exercise 65

Ans: 2.5 cm because 12 - 7 MeV = 5 MeV, hence, range = 5 MeV/2 MeV/cm = 2.5 cm.

CHAPTER 5

Interactions of Photons in Matter

This chapter gives the mathematized account of the many interesting phenomena that happen when light energy in the form of photons interacts with matter. Such phenomena include various forms of scattering and absorption of the photons depending both on the properties of the photons (energy, momentum, intensity) and the properties of matter (elemental composition, mass, charge, etc). The mathematical tools involve formalisms that enable reliable approximations such as time-dependent and time-independent perturbation theory, and Born approximation. Exact theories such as Dirac's relativistic quantum mechanics are also used. Succinctly, elastic scattering (Rayleigh, Thomson), inelastic scattering (Compton), photoelectric absorption of photons by matter and production of electron-positron pairs by photons traversing matter, are treated in this chapter.

5.1 PHOTON ATTENUATION AND THE EXPONENTIAL FUNCTION

Unlike charged particles, a beam of photons is not degraded in energy as it traverses a slap of material but is only attenuated in intensity via several mechanisms such as photoelectric effect, Compton scattering and pair production (see Fig. 5.1). Similar to radioactive decay, the attenuation of photons in matter follows an exponential law. Here are elementary

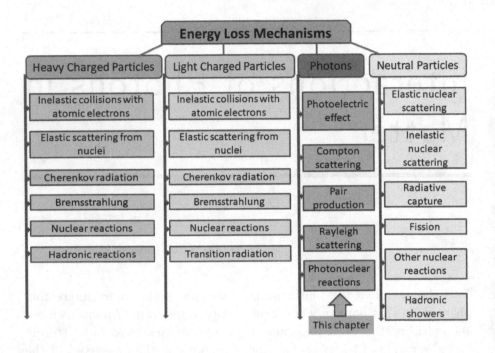

Figure 5.1 Energy Loss Mechanisms for Heavy Charged Particles, Light Charged Particles, Photons and Neutral Projectiles. Properties of the projectiles such as charge, mass, velocity and energy determine the relative predominance of the processes. This chapter deals with photons traversing matter.

mathematical reminders about any exponential function, such as

$$y = ab^x. \tag{5.1}$$

1. When $x > 1$ the function is a growing exponential function.

2. When $x < 0$ the function is a decaying exponential function.

3. All exponential functions have similar shapes because a constant change in the independent variable x gives the same proportional change in the dependent variable y.

4. As the base b increases, the dependent variable's growth rate increases.

5. There is only one value of b such that the gradient of the graph is unity at $x = 0$ and this value of b is denoted e; $e = 2.7182818$ approximately.

6. e can be calculated as a sum of the infinite series $e = \sum_0^\infty \frac{1}{n!} = \frac{1}{0!} + \frac{1}{1!} + \frac{1}{2!} + \dots$

7. e can also be calculated thus: $e = \lim_{x \to \infty} \left(1 + \frac{1}{n}\right)^n$.

8. Owing to above properties of e, the exponential function e^x is crucially important in calculus and physics; for instance, it is the unique nontrivial function (up to multiplication by a constant) which is its own derivative, $\frac{de^x}{dx} = e^x$, and therefore its own antiderivative $\int e^x dx = e^x + C$, etc.

9. The value of e is the base rate of all *continuously* growing or decaying processes

10. e^x and $\ln x$ are inverse expressions: when e^x is amount of continuous growth over a specified time period, then $\ln x$ is the amount of time needed to reach a specified amount of growth, eg, time to double amount of growth is $\ln 2 = 0.693$ and time to reduce to half original amount is $\ln 0.5 = -\ln 2 = -0.693$

When a beam of photons such as X-rays or gamma rays, with intensity I is incident on a material, there is both absorption and scattering of the rays. We describe the fraction of the intensity of photons that is absorbed or scattered per unit thickness dx of the absorber using the **linear attenuation coefficient** μ, usually measured in cm^{-1}. Inter alia, μ will be determined by the probability of being scattered or absorbed. In fact, $\mu = n.\sigma$ where n is the number density of atoms of absorber and σ is the proportionality constant denoting the probability of a photon being scattered or absorbed. As we did in the case of radioactive decay law (Eqs. 1.1a, 1.1b, 1.1c and 1.3), we now consider the change in photon beam intensity at some distance in a material as:

$$dI(x) = -n\sigma I(x)dx. \tag{5.2a}$$

$$\frac{dI}{dx} = -n\sigma I(x). \tag{5.2b}$$

$$\frac{dI}{I(x)} = -n\sigma dx. \tag{5.2c}$$

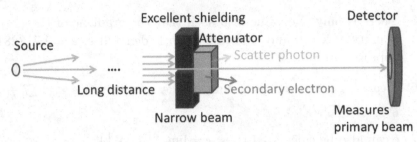

Figure 5.2 Setup for the measurement of photon attenuation through matter using the narrow beam geometry. A narrow beam and long distance are combined to minimize the amount of scatter and secondary particles reaching the detector, so that experiment closely approximates theoretical derivation.

Integrating both sides of Eq. 5.2c, we get $\ln I = -n\sigma x + const$. Now, let $I(x)$ at $x = 0$ be I_0. The constant is thus $\ln I_0$. We then solve for $I(x)$ to obtain

$$I(x) = I_0 e^{-n\sigma x}. \tag{5.3}$$

Eq. 5.3 is the law of photon attenuation through a material, with an equivalent form given by:

$$I(x) = I_0 e^{-\mu x}. \tag{5.4}$$

Thus, photon attenuation through a material is generally an exponential decay, the exponent is negative (see notes 1 to 10 above). The experimental measurement of photon attenuation is illustrated in Fig. 5.2 and this gives meaning to I_0 as the intensity of the photon beam measured when the absorber is removed. There are many excellent text books which give account of the experimental details involved with the setup for measurement of photon attenuation in various materials [3, 23, 44].

Analogous to $N(t) = N_0 e^{-\lambda t}$ which becomes $N(t) = N_0 e^{-(\frac{\ln 2}{T_{1/2}})t}$ for radioactive decay, Eq. 5.4 can be expressed in terms of the distance analog of half-life, namely the *half value layer* or HVL (or $X_{1/2}$) where, μ, the linear attenuation coefficient, is given by:

$$\mu = \frac{\ln 2}{HVL}. \tag{5.5}$$

Thus,

$$I(x) = I_0 e^{-\mu x} = I_0 e^{-(\frac{\ln 2}{HVL})x}. \tag{5.6}$$

Obviously,

$$HVL = \frac{\ln 2}{\mu}. \tag{5.7}$$

Example 5.1 *The HVL of a monoenergetic photon beam in lead is 1 cm. What is the fraction of the original intensity remaining after the beam travels through 1.5 cm of lead?*

Solution

Using Eq. 5.6,

$$I(x) = I_0 e^{-(\frac{\ln 2}{HVL})x} = I_0 e^{-(\frac{0.693}{1})1.5} = 0.3536 I_0.$$

Thus, 35% of I_0 remains.

Example 5.2 *What is the linear attenuation coefficient μ that corresponds to a HVL of 4 cm?*

Solution

Using Eq. 5.7,

$$HVL = \frac{\ln 2}{\mu}.$$

We have,

$$\mu = \frac{\ln 2}{HVL} = \frac{\ln 2}{4\,cm} = \frac{0.693}{4\,cm} = 0.173\ cm^{-1}$$

Exercise 66

What is the linear attenuation coefficient μ that corresponds to a HVL of 5 cm?

Furthermore, just as the mean lifetime or average lifetime of a radioactive nucleus τ is $\frac{1}{\lambda}$, the average distance that an average photon in a beam will travel before interacting, called the mean free path, l, is $\frac{1}{\mu}$ such that we now have

$$I(x) = I_0 e^{-\frac{x}{l}} = I_0 e^{-(\frac{\ln 2}{HVL})x} = I_0 e^{-\frac{1}{1.44 HVL}x}. \tag{5.8}$$

Example 5.3 *What is the mean free path l for an average photon in a beam with $\mu = 0.055/cm$?*

Solution

$$l = \frac{1}{\mu} = \frac{1}{0.055/cm} = 18.18\,cm.$$

Example 5.4 *For purposes of shielding, the so-called tenth value layer TVL, becomes more important than the HVL. Derive a relationship for TVL and HVL.*

Solution

Just as

$$HVL = \frac{\ln 2}{\mu},$$

$$TVL = \frac{\ln 10}{\mu}$$

and so

$$\mu = \frac{\ln 2}{HVL} = \frac{1}{l} = \frac{\ln 10}{TVL}. \tag{5.9}$$

Thus,

$$HVL = (\ln 2) \times l = \left(\frac{\ln 2}{\ln 10}\right) \times TVL \approx (0.301)TVL. \tag{5.10}$$

Since the linear attenuation coefficient $\mu = n\sigma$, and n is number of atoms of absorber per cubic cm, μ depends on the density of the absorber due to n. It is convenient to eliminate this dependence by using **mass attenuation coefficient**, $\frac{\mu}{\rho}$, measured in cm^2 g^{-1}. The mass density ρ used here is measured in g/cm^3. Physically, it is a variable. And we allow it to be mathematically variable by introducing the mass thickness of the absorber layer t in units of g cm^{-2} and recasting Eq. 5.2c as:

$$\frac{dI}{I(x)} = -\left(\frac{\mu}{\rho}\right)\sigma dx. \tag{5.11}$$

The transmitted intensity thus becomes

$$I(t) = I_0 e^{-\int_0^t \frac{\mu}{\rho}(x)dx}. \tag{5.12}$$

For an absorber with constant density, that is, a homogeneous medium,

Eq. 5.12 reduces to the well-known Lambert (1760)–Beer (1852) exponential attenuation law:

$$I(t) = I_0 e^{-\frac{\mu}{\rho}t}.\tag{5.13}$$

How is $\frac{\mu}{\rho}$ obtained *experimentally*? Well, the setup is shown in Fig. 5.2 where a narrow beam of monoenergetic photons with incident intensity I_0, penetrates a layer of the absorber of mass thickness t and emerges with attenuated intensity $I(t)$ picked up by the detector and rewriting Eq. 5.13, one obtains from the measurement:

$$\frac{\mu}{\rho} = \frac{1}{t} \ln \frac{I_0}{I(t)}.\tag{5.14}$$

For experiments and calculations, the $\frac{\mu}{\rho}$ in Eq. 5.13 is in fact the total mass attenuation coefficient which can be expressed in terms of the total atomic cross sections σ_{tot} (σ_{tot} is measured in units of cm^2/atom though units of barns/atom or b/atom are still in use, where 1 barn $= 10^{-24}$ cm^2). Thus, $\frac{\mu}{\rho}$ can have units $(cm^2 g^{-1})$, $(m^2 kg^{-1})$ or other equivalents. To obtain $\frac{\mu}{\rho}$ *theoretically* from total cross section per atom, σ_{tot}:

$$\frac{\mu}{\rho}_{tot} (cm^2 g^{-1}) = \frac{N_A}{uA} \sigma_{tot} = \frac{N_A}{M_A} \sigma_{tot}\tag{5.15}$$

where u is the atomic mass unit which is 1/12th the mass of an atom of C-12 nuclide ($1u = 1.66054 \times 10^{-24}$g)), N_A is Avogadro's number, A and M_A are the relative atomic mass of the composite element(s) and the molar mass of the target material respectively. Defined as in Eq. A.6, σ_{tot} is often denoted as total atomic cross section. Now, various processes contribute to the total atomic cross section σ_{tot} including photoelectric effect, incoherent or Compton scattering, coherent or Rayleigh scattering, pair production and triplet production. These enable σ_{tot} to be given thus [50]:

$$\boxed{\sigma_{tot} = \sigma_{pe} + \sigma_{incoh} + \sigma_{coh} + \sigma_{pair} + \sigma_{ph.n} = \sum_j \sigma_j}\tag{5.16}$$

where the subscripts *pe*, *incoh*, *coh*, *pair*, *trip* and *ph.n* respectively denote cross sections for photoelectric, incoherent scattering, coherent scattering, pair production in the nuclear field, triplet production in atomic electron field and nuclear photoelectric effect and these in turn are the j's in the total cross section sum $\sum_j \sigma$. These processes lead to photon intensity attenuation by removing the photon from the beam through absorption or scattering. Since databases are available mainly for targets that are elements, the mass attenuation coefficient for compounds and mixtures are usually calculated using the additivity rule such that:

$$\frac{\mu}{\rho} = \frac{N_A}{M_A} \sum_j \sigma_j = \sum_j (\frac{\mu}{\rho})_j = N_A \sum_j \sum_i \frac{k_i}{A_i} \sigma_j(Z_i), \tag{5.17}$$

where A_i denotes atomic weight, k_i is the mass fraction of the ith constituent element or atomic species. For incident photons with energy E_γ, the *mass-energy transfer coefficient*, $\frac{\mu_{tr}}{\rho}$ of a target of density ρ is given by:

$$\frac{\mu_{tr}}{\rho} = \frac{1}{\rho dx} \frac{dE_{\gamma,tr}}{E_\gamma} = \frac{N_A}{M_A} \sum_j f_j \sigma_j, \tag{5.18}$$

where $dE_{\gamma,tr}$ denotes mean energy transferred to kinetic energy of charged particles by interactions of the incident photons in traversing distance dx in the target, f_j is the ratio of the mean energy transferred to kinetic energy of charged particles in interaction process j of the incident photon. Since a fraction g of the kinetic energy transferred to charged particles by the photons is subsequently lost to radiative processes such as in-flight annihilation, fluorescence and bremsstrahlung, we keep track of the absorbed energy via the *mass-energy absorption coefficient* $\frac{\mu_{en}}{\rho}$ given by:

$$\frac{\mu_{en}}{\rho} = \frac{\mu_{tr}}{\rho}(1-g). \tag{5.19}$$

Note that f_j is about zero for coherent scattering and this removes coherent scattering cross section σ_{coh} from most calculations of μ_{tr} and μ_{en}. Also, $\sigma_{ph.n}$ are usually excluded from tables or compilations of σ_j. Currently, the most widely used and up to date database of various cross sections and attenuation coefficients of photons with various elements, compounds and mixtures (absorbers) are available at NIST XCOM http://physics.nist.gov/PhysRefData/Xcom/html/xcom1.html. The data base uses an advanced computer code called XCOM which calculates the various cross sections and attenuation coefficients based on theoretical and numerical results.

Example 5.5 *Go to the NIST website and use the XCOM code to calculate and plot the mass attenuation coefficients $\frac{\mu}{\rho}$ in units of $(cm^2 g^{-1})$ for photons of energy 0.001 MeV to 100,000 MeV in carbon and lead. Be*

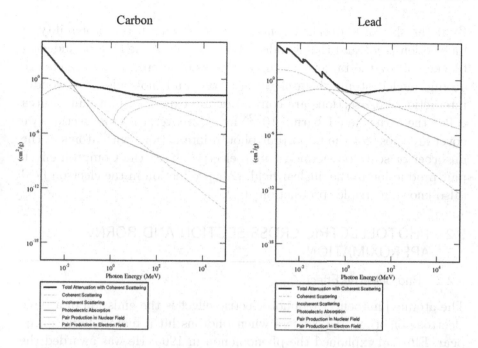

Figure 5.3 Total mass attenuation coefficients for photons in Carbon and Lead, showing contributions from photoelectric absorption, pair production in nuclear field, pair production in electron field (triplet production), coherent scattering and incoherent scattering (Compton) for energies from 1 keV to 100 GeV

sure to show the total and the contributions from photoelectric absorption, pair production in nuclear field, pair production in electron field (triplet production), coherent scattering and incoherent scattering (Compton).

Solution

At NIST XCOM http://physics.nist.gov/PhysRefData/Xcom/html/xcom1.html, type in C or 6 for carbon, Pb or 82 for lead, select required quantities and units, then run the code. The result is shown in Fig. 5.3.

In the following sections, we give the mathematical steps involved in the derivations of these cross sections. Just to ensure physical intuition over the results, we note that the scattering cross section is the effective area for collision. For instance, the cross section σ of a spherical target is πr^2. For a single photon incident on a slab of material of area A, containing one target of cross-sectional area σ, the probability of the photon interacting with the target will be the ratio of the two areas,

$\frac{\sigma}{A}$. If the slab of material A contains n targets, then the probability of interaction is $n\frac{\sigma}{A}$ and this is indeed the fraction of the area occluded or blocked off by the targets. In the processes of interests, the targets are atomic nuclei, atoms themselves, electrons and molecules. No wonder interaction cross-sections are conventionally expressed in a unit of area called the barn where 1 barn $= 10^{-28}$ m^2. Moreover, the cross-sections we cover are those for the important photon interactions with atoms of the absorber or scatterer, namely: photoelectric effect, the Compton effect, pair production in the nuclear field, pair production in the electron field (also known as triplet production) [52].

5.2 PHOTOELECTRIC CROSS-SECTION AND BORN APPROXIMATION

5.2.1 Photoelectric Effect

The atomic photoeffect or photoelectric effect is the emission of atomic electrons or other free carriers when photons hit a material and disappear. Einstein explained the phenomenon in 1905. He was awarded the Nobel Prize in 1921 for "his discovery of the law of the photoelectric effect". The photoelectric effect is basically an interaction mechanism between photons and atomic electrons. Einstein explained that one incident photon with energy E_γ is absorbed by the atom, leading to the emission of one electron with energy E_e, depending on the binding energy of the atom, E_B, or in the case of metals, the *work function*, Φ. Thus, for metallic absorbers:

$$E_e = E_\gamma - \Phi. \tag{5.20}$$

For calculations of photon energies, wavelengths and frequencies, the following Planck-Einstein relation and conversion factors are important:

$$E = hf = \frac{hc}{\lambda} = \frac{1241.5 \text{ eV} \cdot \text{nm}}{\lambda} \approx \frac{1240 \text{ eV} \cdot \text{nm}}{\lambda} \tag{5.21}$$

Example 5.6 *What is the work function Φ for a metal with a threshold wavelength $\lambda_0 = 288\,nm$?*

Solution

Applying Eq. 5.21 to the case of Φ.

$$\Phi = hf_0 = \frac{hc}{\lambda_0} = \frac{1240 \text{ eV} \cdot \text{nm}}{\lambda_0} \tag{5.22}$$

Thus,

$$\Phi = \frac{1240 \text{ eV} \cdot \text{nm}}{\lambda_0} = \frac{1240 \text{ eV} \cdot \text{nm}}{288 \text{ nm}} = 4.31 \text{ eV}.$$

Example 5.7 *The binding energy of the K-shell of Tungsten or Wolfram (W) $E_B(K)$ is 69.525 keV. A photon of energy 100 keV is incident on W and disappears, producing a photoelectron. (a) What is the energy of the photoelectron E_e? (b) Give an expression for the residual energy of the W ion or of any such ion, E_{A+} even though this residual energy is negligible.*

Solution

(a) Eq. 5.20 can also be expressed more accurately in terms of the energy available for transfer to both photoelectron and ionized atom of absorber, E_{tr}:

$$E_{tr} = E_\gamma - E_B = E_e + E_{A+}. \tag{5.23}$$

Thus,

$$E_{tr} = 100 \text{ keV} - 69.525 \text{ keV} = 30.475 \text{ keV}.$$

E_e will be about 30 keV, based on the negligibility of E_{A+}. (b) Using ratios of masses involved, the exact kinetic energy of photoelectron is obtained as follows:

$$E_{tr} = E_e + E_{A+} = \frac{M_{A+}}{M_{A+} + m_e} E_{tr} + \frac{m_e}{M_{A+} + m_e} E_{tr}.$$

Likewise,

$$E_{tr} = E_e + E_{A+} = \frac{M_{A+}}{M_{A+} + m_e}(E_\gamma - E_B) + \frac{m_e}{M_{A+} + m_e}(E_\gamma - E_B).$$

Since the mass of the un-ionized or parent atom, M_A, is $M_{A+} + m_e$,

$$E_e = \frac{M_{A+}}{M_A}(E_\gamma - E_B). \tag{5.24}$$

The exact kinetic energy of residual ion is:

$$E_{A+} = \frac{m_e}{M_A}(E_\gamma - E_B). \tag{5.25}$$

Since $\frac{m_e}{M_A}$ for the mass-energy of the electron and mass-energy of W atom gives

$$\frac{0.511 \text{ MeV}}{931.5 \text{ MeV/amu} \times 183.84 \text{ amu}} = 2.98 \times 10^{-6},$$

we see that E_{A+} is truly negligible.

Example 5.8 *What is the binding energy ε_b for K-shell electrons, in terms of the Rydberg energy R_y? or Rydberg constant.*

Solution

The binding energy ε_b for K-shell electrons is given by:

$$\varepsilon_b(K) \approx R_y(Z-1)^2 \tag{5.26}$$

where $R_y = 13.61$ eV.

5.2.2 Photoelectric Cross Section from Perturbation Theory

The photoelectric cross section σ_{pe} gives the probability of the photo-effect occurring [25, 62]. A rule of thumb approximation gives σ_{pe} for photon energies E above the highest atomic binding energy:

$$\sigma_{pe} \approx c \cdot \frac{Z^n}{E^3}, \tag{5.27}$$

where c is a constant and n varies between 4 and 5. Between 0.1 and 3 MeV, n varies from 4 to 4.6, with the energy exponent decreasing from 3 to 1 when $E \geq m_e c^2$. Clearly, the higher the photon energy, the lower the probability of photoelectric effect occurring. Likewise, high atomic number materials such as lead ($Z = 82$), produce more photoelectric interactions than low atomic number materials. Hence, lead is the preferred material for photon shielding in NIM labs, radiotherapy units, nuclear reactors, etc. The derivation of the atomic photoelectric cross section σ_{pe} involves wavefunctions for many atomic electrons. For a detailed work on the derivation with extensive mathematical details, see recent work by Sabbatucci and Salvat [103] where atomic states are described within the independent-electron approximation. Moreover, bound and free one-electron orbitals are obtained as solutions of the Dirac equation with the Dirac-Hartree-Fock-Slater (DHFS) self-consistent potential of the ground-state configuration. We give a few of the mathematical steps involved. First order perturbation theory is used since the interaction between the atom and the electromagnetic field is weak. The one-electron Dirac wave equation is solved for a central potential $V(r)$, the solutions being one electron orbitals Ψ, representing states of the target atoms via so-called Slatter determinants. Following absorption of the photon, the active electron is considered to jump from a bound orbital Ψ_a of energy ε_a to a final orbital Ψ_b with energy ε_b such that:

$$\varepsilon_b = \varepsilon_a + E_\gamma. \tag{5.28}$$

Eq. 5.28 ignores the kinetic energy of the recoiling atom since $m_e/M_{atom} \approx 10^{-4}$. For a beam of photons with energy $E_\gamma = \hbar\omega$, the cross section for excitation of the active electron to a bound orbital, Ψ_b is given by:

$$\sigma_{pe} = \frac{(2\pi)^2 e^2 c\hbar}{E_\gamma} |M|^2 \delta(\varepsilon_b - \varepsilon_a - E_\gamma), \tag{5.29}$$

where M is the transition matrix with elements that include the electron orbitals such that

$$M \approx \langle \Psi_b | |\alpha e^{i\mathbf{k \cdot r}}| |\Psi_a \rangle, \tag{5.30}$$

where $\mathbf{k}$ is the wave vector. Calculations lead to approximations which give the photoelectric cross section in Eq. 5.27. Most early calculations of the atomic photoeffect were for the K-shell only but recent work has included all sub-shells for elements $Z = 1$ to $Z = 101$. Heitler [47] derived σ_{pe} above the K- absorption edge using non-relativistic Born approximation within first order pertubation theory (see Appendix A.3). A sketch of the mathematical physics steps he used is as follows.

5.2.2.1 Initial and Final States

The initial state consists of two particles, namely, a bound electron (atomic electron) with approximate wavefunction

$$\Psi_a \approx e^{\frac{-r}{a}}$$

with $a = \frac{a_0}{Z}$, $a_0 = \frac{\hbar^2}{m_e e^2}$ being the Bohr radius; as well as an incident photon with free particle wave function

$$\Psi_\gamma \approx e^{i\vec{k}.\vec{r}}.$$

The final state is just a free electron with wavefunction given by

$$\Psi_b \approx e^{i\frac{\vec{p}.\vec{r}}{\hbar}}.$$

5.2.2.2 Interaction and Result

Expressing the interaction potential as

$$V = \vec{A}.\vec{p}$$

where $\vec{A}$ represents the vector potential of the electromagnetic radiation or photon and $\vec{p}$ the electron momentum. Heitler's first order perturbation calculation in the Born approximation yielded the following differential cross-section for the atomic photoelectric effect:

$$\frac{d\sigma_{pe}}{d\Omega} = 4\sqrt{2}\,\frac{r_e^2 Z^5}{(137)^4}\left(\frac{m_e c^2}{\hbar\omega}\right)^{7/2}\left[\frac{(\sin^2\theta\cos^2\phi)^2}{(1-\frac{v}{c}\cos\theta)^4}\right], \qquad (5.31)$$

where θ and ϕ are polar and azimuthal angles which specify the direction of the photoelectron.

Exercise 67

Using integration over solid angle, obtain the total photoelectric cross section from Eq. 5.31. Compare your result with Heitler's:

$$\int \frac{d\sigma_{pe}}{d\Omega}d\Omega = 4\sqrt{2}\,\frac{r_e^2 Z^5}{(137)^4}\left(\frac{m_e c^2}{\hbar\omega}\right)^{7/2}\sigma_0 \qquad (5.32)$$

where Heitler ignored the angular dependence in the denominator and then accounted for two electrons in the K-shell by doubling the result. Note that σ_0 is the Thomson electronic cross section which we derive below (see Eq. 5.93).

The photoelectric absorption curve shows absorption edges (see Fig. 5.3) whenever incoming photon energy $h\nu$ matches the ionization energy of K, L, M,... shells of atoms. These shells can have substructures. We note that the photoelectric absorption probability is larger for more tightly bound electrons which are found in the K-shell because unbound electrons cannot absorb photons without re-emitting electromagnetic radiation. Note that the most widely accepted and up to date source for the photoelectric cross sections or attenuation coefficients of photons with various elements, compounds and mixtures (absorbers) are available at NIST XCOM http://physics.nist.gov/PhysRefData/Xcom/html/xcom1.html.

Example 5.9 *Owing to the ubiquity of the Born approximation for scattering problems in nuclear physics [53, 63] and beyond, illustrate and state in general terms the Born approximation for the scattering factor f for a potential $V(r)$.*

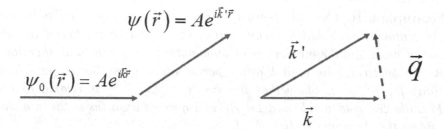

$$\psi(\vec{r}) = Ae^{i\vec{k}'\cdot\vec{r}}$$

$$\psi_0(\vec{r}) = Ae^{i\vec{k}\cdot\vec{r}}$$

Figure 5.4 First order Born approximation where a plane wave $\psi_0(r)$ is scattered into a wave of same amplitude, $\psi(r)$ as the incoming wave but with different direction.

Solution

As shown in Fig. 5.4, the incoming wave is considered a plane wave of direction $\vec{k}$ which upon interaction gets deflected in a direction $\vec{k}'$ but keeps same amplitude and same wavelength. The scattering vector $\vec{q}$ is given by $\vec{k}' - \vec{k}$ and its magnitude is

$$q = 2k\sin\left(\frac{\theta}{2}\right) = \frac{4\pi}{\lambda}\sin\left(\frac{\theta}{2}\right). \tag{5.33}$$

Using $\vec{q}$ as defined and with the physical meaning of momentum transfer, the scattering factor for the Born first approximation is given by

$$f_{Born}(\theta,\phi) = \frac{-2m}{\hbar^2}\frac{1}{4\pi}\int e^{i\vec{q}\cdot\vec{r}}V(\vec{r})d^3\vec{r}. \tag{5.34}$$

Note that the scattering factor or scattering amplitude f_{Born} is proportional to the Fourier transform of the potential $V(r)$ with respect to q [90]. Details on Born Approximation, First Born Approximation, Born Series, Fermi's Golden Rule, etc., can be found in Appendix A.3. Analytic expressions for $\frac{d\sigma_{pe}}{d\Omega}$ for ionization of the K shell electrons of atoms by linearly polarized photons derived by Sauter [104] using the Born approximation are still in use even in Monte Carlo codes for calculations of photon transport [57, 103]. Note that in making the Born approximation, Sauter replaced the plane waves we stated above with the so-called Dirac plane waves to stay in the context of relativistic quantum mechanics. Thus, there exists a relativistic plane-wave Born approximation.

Example 5.10 *Current theory of the photoelectric effect builds on the elementary theory which assumes that the interaction between the absorber atom and the incident electromagnetic field is weak and, therefore, it can be treated as a first order perturbation (first Born approximation)* [7, 103, 104]. *Show that the result in Eq. 5.29 is related to the Fermi's 2nd golden rule and 1st Born Approximation (as outlined above and in the Appendix section A.3).*

Solution

Let the bound orbital to which the electron is excited be represented by

$$\psi_b = \psi_{n_b \kappa_b m_b}$$

where n, κ and m denote principal, relativistic and magnetic (spin) quantum numbers respectively. In the case of ionization, the final orbital b belongs to a continuum of states with energy spectrum $\varepsilon_b > 0$. We rewrite Eq. 5.29 explicitly as the cross section for a bound electron to be excited from ψ_a to ψ_b:

$$\sigma_{n_b \kappa_b m_b : n_a \kappa_a m_a} = \frac{(2\pi)^2 e^2 c\hbar}{E_\gamma} |M_{n_b \kappa_b m_b : n_a \kappa_a m_a}|^2 \delta(\varepsilon_b - \varepsilon_a - E_\gamma), \quad (5.35)$$

where M is the transition matrix with elements that include the electron orbitals such that

$$M_{n_b \kappa_b m_b : n_a \kappa_a m_a} \approx \langle \Psi_{n_b \kappa_b m_b} | |\alpha e^{i\mathbf{k}\cdot\mathbf{r}}| |\Psi_{n_a \kappa_a m_a} \rangle. \quad (5.36)$$

In Eq. 5.35, the delta distribution accounts for the fact that absorption happens only at resonance, when the excitation energy $\varepsilon_b - \varepsilon_a$ coincides with incident photon energy E_γ. Then applying Fermi's 2nd golden rule (see outline in section A.3) the differential cross section for absorption of a photon, with emission of the active electron with linear momentum $\hbar k_b$ in the direction within solid angle $d\Omega$ becomes [94, 103, 107]:

$$\frac{d\sigma_{n_b \kappa_b m_b : n_a \kappa_a m_a}}{d\Omega}$$
$$= \frac{(2\pi)^2 e^2}{E_\gamma} \frac{k_b(\varepsilon_b + m_e c^2)}{c\hbar} |M_{n_b \kappa_b m_b : n_a \kappa_a m_a}|^2 \delta(\varepsilon_b - \varepsilon_a - E_\gamma), \quad (5.37)$$

where M is the transition matrix with elements:

$$M_{n_b \kappa_b m_b : n_a \kappa_a m_a} \approx \langle \Psi_{n_b \kappa_b m_b} | |\alpha e^{i\mathbf{k}\cdot\mathbf{r}}| |\Psi_{n_a \kappa_a m_a} \rangle, \quad (5.38)$$

is evaluated on the energy shell $\varepsilon_b = \varepsilon_a + E_\gamma$. Note that for ionization, the final orbital belongs to the continuum spectrum where $\varepsilon_b > 0$.

Example 5.11 *For both excitation and ionization by low energy photons such that the wavelength $\lambda = \frac{2\pi}{k}$ is much greater than the mean radial distance of electrons in the active subshell, show that the calculation of the differential cross section can be simplified using the dipole approximation.*

Solution

In the dipole approximation, $\vec{k}.\vec{r} << 1$, hence, the exponential in the transition matrix elements M of Eq. 5.30 and Eq. 5.38 is substituted with unity:

$$M_{n_b \kappa_b m_b : n_a \kappa_a m_a} \approx \langle \Psi_{n_b \kappa_b m_b} | |\alpha| |\Psi_{n_a \kappa_a m_a} \rangle, \qquad (5.39)$$

making the final calculation much simpler. Note that even α in Eq. 5.39 can now be replaced with just $\vec{r}$ using the commutation relation of the Dirac Hamiltonian (see Section 5.3, where we give the Dirac equation in the context of the derivation of Klein-Nishina formula for Compton Scattering).

Example 5.12 *As already noted, in Eq. 5.35, the delta distribution mathematically accounts for the fact that photoelectric absorption happens only at resonance, when the excitation energy $\varepsilon_b - \varepsilon_a$ coincides with incident photon energy E_γ. Physically this leads to the observation of absorption edges in the photoelectric cross section (Fig. 5.3 for Pb showing edges since $Z = 82$ but less obvious edges for C with $Z = 6$). Let the contribution of photoelectric effect to the total mass attenuation coefficient $\frac{\mu}{\rho_{tot}}$ be $\frac{\mu_{pe}}{\rho}$, such that*

$$\frac{\mu_{pe}}{\rho}(\text{cm}^2\text{g}^{-1}) = \frac{N_A}{uA}\sigma_{pe} = \frac{N_A}{M_A}\sigma_{pe}. \qquad (5.40)$$

Using Eq. 5.40, derive an expression for the probability P_i^{pe} of the photoelectric effect occurring within a subshell i of an absorber, when $\frac{\mu_{pe}}{\rho}$ is plotted against E_γ in the range including absorption edges such as K, L and M.

Solution

For any subshell i, we define an absorption edge parameter A_i:

$$A_i = \frac{(\frac{\mu_{pe}}{\rho})^H - (\frac{\mu_{pe}}{\rho})^L}{(\frac{\mu_{pe}}{\rho})^H}, \qquad (5.41)$$

where H and L are the high and low values of the photoelectric mass attenuation coefficient $\frac{\mu_{pe}}{\rho}$ at the absorption edge i under consideration. For total probability $\Sigma P_i = 1$, then

$$P_i = \left(1 - \sum_{n=0}^{i-1} P_n\right) A_i, \qquad (5.42)$$

where $P_{n=0} = 0$.

Example 5.13 *(a) Use the same procedure in Fig. 5.3 to obtain XCOM data (at NIST XCOM) for lead between 1 keV and 100 keV. Focusing on the photoelectric mass attenuation coefficient $\frac{\mu_{pe}}{\rho}$ as a function of energy, note the energies for absorption edges at K, L_1, L_2 and L_3 and the high and low values of $\frac{\mu_{pe}}{\rho}$ for these energies. (b). Determine the probabilities that if the photoelectric effect occurs, it will occur in the K-shell, P_K and L-shell P_L, for incident photon with energy greater than the K-shell binding energy: $E_\gamma \geq E_B(K)$.*

Solution

(a) From the table or graph generated as required, we get:

$$E_B(K) = 88 \text{ keV}, \quad (\frac{\mu_{pe}}{\rho})^H = 7.321 \text{ cm}^2\text{g}^{-1}, \quad (\frac{\mu_{pe}}{\rho})^L = 1.547 \text{ cm}^2\text{g}^{-1}$$

$$E_B(L_1) = 15.86 \text{ keV}, \quad (\frac{\mu_{pe}}{\rho})^H = 151.7 \text{ cm}^2\text{g}^{-1}, \quad (\frac{\mu_{pe}}{\rho})^L = 131.2 \text{ cm}^2\text{g}^{-1}$$

$$E_B(L_2) = 15.2 \text{ keV}, \quad (\frac{\mu_{pe}}{\rho})^H = 145.2 \text{ cm}^2\text{g}^{-1}, \quad (\frac{\mu_{pe}}{\rho})^L = 104.5 \text{ cm}^2\text{g}^{-1}$$

$$E_B(L_3) = 13.04 \text{ keV}, \quad (\frac{\mu_{pe}}{\rho})^H = 158.2 \text{ cm}^2\text{g}^{-1}, \quad (\frac{\mu_{pe}}{\rho})^L = 63.1 \text{ cm}^2\text{g}^{-1}$$

(b) Using Eq. 5.42 and Eq. 5.42,

$$P_K = \left(1 - \sum_{n=0}^{0} P_0\right) A_K = A_K = \frac{7.321 - 1.547}{7.321} = 0.789$$

Hence, if the photoelectric effect occurs, there is a 79% chance of occurring in the K-shell.

$$P_{L_1} = \left(1 - \sum_{n=0}^{1} P_i\right)A_{L_1} = (1 - P_k)A_{L_1}$$

$$= (1 - 0.789) \times \frac{151.7 - 131.2}{151.7} = 0.0285$$

Hence, if the photoelectric effect occurs, there is a 2.9% chance of occurring in the L_1-shell.

$$P_{L_2} = \left(1 - \sum_{n=0}^{2} P_i\right)A_{L_2} = (1 - P_k - P_{L_1})A_{L_2}$$

$$= (1 - 0.789 - 0.0285) \times \frac{145.2 - 104.5}{145.2} = 0.0512$$

Hence, if the photoelectric effect occurs, there is a 5.1% chance of occurring in the L_2-shell. Finally,

$$P_{L_3} = \left(1 - \sum_{n=0}^{3} P_i\right)A_{L_3} = (1 - P_k - P_{L_1} - P_{L_2})A_{L_3}$$

$$= (1 - 0.789 - 0.0285 - 0.0512) \times \frac{158.2 - 63.1}{158.2} = 0.0789$$

Hence, if the photoelectric effect occurs, there is a 7.9% chance of occurring in the L_3-shell. Of course, probability occuring in L-shell is

$$P_L = P_{L_1} + P_{L_2} + P_{L_3} = 0.1586$$

or 15.9%.

Exercise 68

(a) Use the same procedure in Fig. 5.3 to obtain XCOM data (at NIST XCOM) for Tungsten between 1 keV and 100 keV. Focusing on the photoelectric mass attenuation coefficient $\frac{\mu_{pe}}{\rho}$ as a function of energy, note the energies for absorption edges at K, L_1, L_2 and L_3. (b) Determine the probabilities that *if the photoelectric effect occurs*, it will occur in the K-shell, P_K and L-shell P_L, for incident photon with energy greater than the binding K-shell binding energy: $E_\gamma \geq E_B(K)$. You will need the high and low values of $\frac{\mu_{pe}}{\rho}$ for the energies in (a). (c) How would the result in (b) be different if the incident photon energy were less than the K-shell binding energy but greater than or equal to the L_1-shell binding energy?

5.3 KLEIN-NISHINA FORMULA FOR COMPTON SCATTERING

5.3.1 Compton Scattering

The kinematics of Compton scattering is well known and easy to illustrate. The title of his work describing the phenomenon is instructive: "A quantum theory of the scattering of X-rays by light elements" [28]. Compton scattering is a quantum process involving energy and momentum transfers between a single photon and a free electron. Compton discovered the effect in 1922 and won the Nobel prize in 1927 "for his discovery of the phenomenon named after him—the Compton Effect". Specifically, Compton scattering is inelastic scattering of photons by charged particles in which some energy and momentum are transferred to the charged particle while the photon moves off with a reduced energy and change(s) in momentum (direction and magnitude). Since photons are mass-less relativistic particles and since the energy transferred to the electron is comparable to its rest energy, relativistic values of the energy and momenta are used in the theory of Compton scattering as follows. Einstein's special theory of relativity gives the relativistic mass m of an object moving with velocity v and having rest mass m_0 as:

$$m = \frac{m_0}{\sqrt{1 - (\frac{v}{c})^2}} = \frac{m_0}{\sqrt{1 - (\beta)^2}} = \gamma m_0. \tag{5.43}$$

The relativistic momentum can be expressed in terms of this relativistic mass, thus: $\vec{p} = m\vec{v}$. Hence, squaring both sides of Eq 5.43 and multiplying by c^4 gives:

$$m^2 c^4 - m^2 v^2 c^2 = m_0^2 c^4, \tag{5.44}$$

$$(mc^2)^2 - (pc)^2 = (m_0 c^2)^2, \tag{5.45}$$

$$E^2 - (pc)^2 = E_0^2, \tag{5.46}$$

$$(pc)^2 = E^2 - E_0^2, \tag{5.47}$$

which relates the magnitude of the relativistic momentum $\vec{p}$ of an object with its relativistic energy E and rest energy E_0. Thus for a photon,

$$pc = E, \tag{5.48}$$

$$p = \frac{E}{c} = \frac{h\nu}{c} = \frac{h}{\lambda}. \tag{5.49}$$

Compton applied conservation of energy and momentum to these

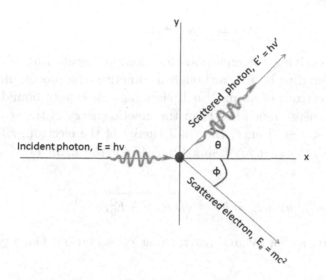

Figure 5.5 Compton Scattering. An incident photon scatters inelastically from an atomic electron with a change in frequency and wavelength.

relativistic expressions along with quantized energy formula for photons and arrived at his theory for the inelastic scattering of photon and elcctron. We next recapitulate the essential derivations and results.

Consider the energy $E = h\nu$ and momentum $p = \frac{h\nu}{c}\hat{x}$ of a photon propagating in the positive x direction. Let the frequency and propagating direction of the photon before scattering off an electron be ν and $\hat{x}$ respectively. After the scattering, these become ν' and $\hat{x}'$. Using simple Cartesian coordinates, the photon is scattered in the positive x and y directions at an angle θ (see Fig 5.5). The magnitude of momentum of scattered photon is $p' = \frac{h\nu'}{c}$ and its energy is $E' = h\nu'$.

The momentum of the electron before the scattering $p_e = 0$ because the electron is considered to be at rest so its velocity is 0. After the scattering, electron momentum is $p'_e = \frac{1}{c}\sqrt{E_e^2 - E_0^2}$ where the electron energy $E_e = mc^2$ and m is the relativistic mass of the electron. The electron is scattered in the positive x and negative y directions at an angle ϕ with respect to the positive x-direction (see Fig. 5.5). The conservation of energy requires that the energy of the incident gamma ray, E, and the rest energy of the electron E_o before scattering be equal to the energy of the scattered gamma ray, E' and the total energy of the electron, E_e, after scattering:

$$h\nu + E_0 = h\nu' + E_e. \tag{5.50}$$

Note that in deriving the equations for Compton scattering, the electron is considered to be free although in practice, the process involves the valence electrons of atoms. Such electrons are loosely bound, with their binding energy much less than the kinetic energy of the scattered electron: $E_b \ll E_{KE}$. Hence, the total energy of the electron, E_e, after scattering is $E_e = E_{KE} + E_b$. Using Eq. 5.47 for the E_e in Eq. 5.50, we obtain,

$$h\nu + E_0 = h\nu' + \sqrt{(p_e'c)^2 + E_0^2}. \tag{5.51}$$

Using pc for $h\nu$, $p'c$ for $h\nu'$ and rearranging, conservation of energy leads to:

$$pc + E_0 = p'c + \sqrt{(p_e'c)^2 + E_0^2}, \tag{5.52}$$

$$(p - p')c + E_0 = \sqrt{(p_e'c)^2 + E_0^2}, \tag{5.53}$$

$$(p - p')^2 c^2 + E_0^2 = (p_e'c)^2 + E_0^2, \tag{5.54}$$

$$p^2 + p'^2 - 2pp' + \frac{2(p - p')E_0}{c} = p_e'^2. \tag{5.55}$$

Now, conservation of energy has yielded an equation for the momentum of the electron p_e. Application of conservation of momentum yields a second independent expression for p_e. Using Cartesian coordinates shown (Fig 5.5) we have, for the x and y components of the momenta respectively,

$$p = p'\cos\theta + p_e'\cos\phi, \tag{5.56}$$

$$0 = p'\sin\theta - p_e'\sin\phi. \tag{5.57}$$

Rearranging the x- and y-momentum equations and squaring both sides yield:

$$p^2 + p'^2\cos^2\theta = p_e'^2\cos^2\phi, \tag{5.58}$$

$$p'^2\sin^2\theta = p_e'^2\sin^2\phi. \tag{5.59}$$

Using the identity $\cos^2 x + \sin^2 x = 1$, x- and y- squared momenta simplify to:

$$p^2 + p'^2 - 2pp' \cos\theta = p_e'^2. \tag{5.60}$$

Equating Eq. 5.55 and Eq. 5.60, we have:

$$p^2 + p'^2 - 2pp' + \frac{2(p - p')E_0}{c} = p^2 + p'^2 - 2pp' \cos\theta. \tag{5.61}$$

Eq. 5.61 reduces to

$$\frac{1}{p'} - \frac{1}{p} = \frac{c}{E_0}(1 - \cos\theta). \tag{5.62}$$

Also, using Eq. 5.49, Eq. 5.62 can be rendered as:

$$\frac{1}{E'} - \frac{1}{E} = \frac{1}{E_0}(1 - \cos\theta). \tag{5.63}$$

Now, inverting and solving for the energy of scattered photon, we have the familiar or convenient expression for Compton scattering:

$$\boxed{E' = \frac{E}{1 + \frac{E}{m_0 c^2}(1 - \cos\theta)},} \tag{5.64}$$

or with respect to changed frequencies:

$$\boxed{h\nu' = \frac{h\nu}{1 + \frac{h\nu}{m_0 c^2}(1 - \cos\theta)}.} \tag{5.65}$$

Note that $E_o = m_o c^2$ is the electron rest mass or rest energy. Also, for quick comparisons and calculations, Eq. 5.65 can be rendered as:

$$\boxed{h\nu' = \frac{h\nu}{1 + \alpha(1 - \cos\theta)},} \tag{5.66}$$

where $\alpha = \frac{h\nu}{m_0 c^2}$. Obviously, conservation of energy can be applied to solve for the energy of the electron.

Example 5.14 *Gamma rays with maximum energy 2.5 MeV are emitted from a radioactive nucleus and interact with atomic electrons. What is the maximum energy of the back-scattered photons?.*

Solution

We use Eq. 5.65:

$$hv' = \frac{hv}{1 + \frac{hv}{m_0 c^2}(1 - \cos\theta)}.$$

The maximum energy of back-scattered photons is obtained when $\theta = 180°$. Hence,

$$hv' = \frac{hv}{1 + \frac{hv}{m_0 c^2}(1 - \cos 180)} = \frac{hv}{1 + \frac{2hv}{m_0 c^2}} = \frac{2.5}{1 + \frac{5}{0.511}} = 0.23 \text{ MeV}.$$

Exercise 69

A 10 MeV photon from a medical linear accelerator (linac) collides with an electron in the tissue of a patient. What is the energy of the photon after it scatters at a right angle? Hint: use Eq. 5.65 with $\theta = 90°$.

Exercise 70

(a) What is the kinetic energy of a Compton electron after it is scattered by a photon of energy 0.1 MeV, itself scattered by 160° following interaction with the electron? Hint: use Eq. 5.77 with $\theta = 160°$ (b) What is the energy of the scatttered photon?

Example 5.15 *For a 0.662 MeV gamma ray emitted from the decay of the Cs-137 nucleus, what is the maximum energy of the electron emitted from the material of a NaI(Tl) detector? Obtain a general relation for maximum energy of the electron in Compton scattering.*

Solution

Maximum kinetic energy is transferred to the electron during a head-on collision with a photon in which the photon rebounds back at $\theta = 180°$. Hence, using Eq. 5.77, we have

$$E_{eMax} = E - E' = E\left[1 - \frac{1}{1 + \frac{E}{m_0 c^2}(1 - \cos 180°)}\right] \quad (5.67)$$

$$= E\left[1 - \frac{1}{1 + 2\frac{E}{m_0 c^2}}\right]. \quad (5.68)$$

Substituting the values,

$$E_{eMax} = 0.662 \left[1 - \frac{1}{1 + 2\frac{0.662}{0.511}} \right] = 0.477. \quad (5.69)$$

Thus, for a 0.662 MeV gamma photon from Cs-137 decay, the maximum kinetic energy of the electron is 0.477 MeV. Generalizing, the relation for the maximum energy of scattered Compton electrons is

$$\boxed{E_{eMax} = E \left[1 - \frac{1}{1 + 2\frac{E}{m_0 c^2}} \right] = E \left[\frac{2\alpha}{1 + 2\alpha} \right]} \quad (5.70)$$

where $\alpha = \frac{E}{m_0 c^2}$. This maximum energy of the electron gives the so-called *Compton edge* in Compton spectrum. Conversely, the minimum energy carried by the scattered photon (at 180° back scattering) is:

$$E' = \frac{E}{1 + 2\alpha}, \quad (5.71)$$

or with respect to frequency,

$$h\nu' = \frac{h\nu}{1 + 2\alpha}. \quad (5.72)$$

Example 5.16 *What is the change in wavelength of the photon, $\lambda' - \lambda$ due to the Compton effect?*

Solution

$$\frac{1}{E'} - \frac{1}{E} = \frac{1}{E_0}(1 - \cos\theta) \implies \frac{\lambda'}{hc} - \frac{\lambda}{hc} = \frac{1}{m_e c^2}(1 - \cos\theta); \quad (5.73)$$

$$\lambda' - \lambda = \frac{h}{m_e c}(1 - \cos\theta) \quad (5.74)$$

where $\frac{h}{m_e c}$ is the so-called Compton wavelength, λ_c. Numerically, $\lambda_c = 2.43 \times 10^{-12}$m.

Example 5.17 *For the 0.661 MeV gamma photon of Cs-137, back-scattered at 180°, calculate (a) the wavelength of the incident photon, the change in wavelength due to Compton effect and the wavelength of the back-scattered photon; (b) the energy of the back-scattered photon; (c) the energy of the resultant Compton electron*

Solution

(a) First we find the wavelength of the incident gamma ray, using

$$\lambda = \frac{hc}{E} = \frac{1240 \text{ eV nm}}{E} \tag{5.75}$$

which gives

$$\lambda = \frac{1240 \text{ eV nm}}{0.661 \times 10^6 \text{ eV}} = 0.00188 \text{ nm}$$

The change in wavelength is

$$\lambda' - \lambda = \frac{h}{m_e c}(1 - \cos\theta) = \lambda_c = 2.43 \times 10^{-12}(1 - \cos 180°) \text{ m} = 0.00484 \text{ nm}$$

Thus, the wavelength of the back-scattered photon is

$$\lambda' = \lambda + \Delta\lambda = 0.00188 + 0.00484 = 0.00672 \text{ nm}.$$

(b) We can obtain the energy of the scattered photon using Eq. 5.66 or using the scattered wavelength,

$$E' = \frac{1240 \text{ eV nm}}{\lambda} = \frac{1240}{0.00672} = 0.185 \text{ MeV}.$$

(c) For the resultant Compton electron, it is straightforward at this point to obtain its maximum energy as the difference between incident and scattered photon energies (Eq. 5.77), thus:

$$E - E' = 0.661 - 0.185 \text{ MeV} = 0.467 \text{ MeV}.$$

Exercise 71

Show, using the principle of conservation of momentum, that the recoil momentum of the scattered electron is equal to the vector difference between the momenta of the incident photon and scattered photon. That, is, show explicitly that

$$\left(\frac{m\beta c}{\sqrt{1-\beta^2}}\right)^2 = \left(\frac{h\nu}{c}\right)^2 + \left(\frac{h\nu'}{c}\right)^2 - 2\frac{h\nu}{c} \cdot \frac{h\nu'}{c} \cdot \cos\theta. \tag{5.76}$$

Exercise 72

Using the principle of conservation of energy, that is, starting from Eq. 5.50, show that the energy of the scattered electron E_e is given by

$$E_e = E - E' = E\left[1 - \frac{1}{1 + \frac{E}{m_0 c^2}(1 - \cos\theta)}\right] \tag{5.77}$$

Exercise 73

Using Eq. 5.77 and the principles of conservation of energy and momentum, show that the angle for the scattered photon θ and the angle for the scattered (recoil) electron ϕ are related thus:

$$\cos \theta = \left[1 - \frac{2}{(1+\alpha)^2 \tan^2 \phi + 1} \right] \tag{5.78}$$

$$\cot \phi = (1+\alpha) \tan \frac{\theta}{2}, \tag{5.79}$$

where $\alpha = \frac{h\nu}{m_0 c^2}$. Note that although the scattering angle for the photon θ can vary from 0 to 180°, the scattering angle for the electron only varies from 0 to 90° with respect to the direction of the incoming photon.

Exercise 74

The electron threshold energy for a Cerenkov counter whose index of refraction is 1.6 is found to be 0.143 MeV (see Eq. 4.44). What gamma photon energy interacting via Compton Scattering would produce this electron with energy 0.143 MeV? Hint: 180°back-scattered photon would transfer the most energy to the electron. Hence, use Eq. 5.69.

5.3.2   Derivation of the Klein-Nishina formula

To enable the calculation of the probability of photons scattering at certain angles and imparting energy E_e to the electron, Oskar Klein and Yoshio Nishina [59, 60] derived the cross section for Compton or incoherent scattering, assuming free electron and thereby using Born approximation. They based their derivation on the then newly available Dirac equation for the electron [30, 31], a formula that accounts for relativity and quantum electrodynamics. They started with the Dirac equation,

$$i\hbar \frac{\partial \Psi(r,t)}{\partial t} = c \left(\frac{\hbar}{i} \vec{\alpha} \cdot \vec{\nabla} + \beta mc \right) \Psi(r,t) \tag{5.80}$$

or in Dirac notation,

$$i\hbar \frac{\partial \left| \Psi(r,t) \right\rangle}{\partial t} = c \left(\vec{\alpha} \cdot \vec{P} + \beta mc \right) \left| \Psi(r,t) \right\rangle, \tag{5.81}$$

where $\vec{\alpha}$ and β are 4×4 anti-commuting Hermitian matrices:

$$\alpha_j = \begin{pmatrix} 0 & \sigma_j \\ \sigma_j & 0 \end{pmatrix}$$

and:

$$\beta = \begin{pmatrix} I & 0 \\ 0 & -I \end{pmatrix}$$

with $I = \begin{pmatrix} 1 & 0 \\ 0 & 1 \end{pmatrix}$, $\sigma_x = \begin{pmatrix} 0 & 1 \\ 1 & 0 \end{pmatrix}$, $\sigma_y = \begin{pmatrix} 0 & -i \\ i & 0 \end{pmatrix}$, $\sigma_z = \begin{pmatrix} 1 & 0 \\ 0 & -1 \end{pmatrix}$.

Since the Dirac equation considers both relativity and quantum electrodynamics, leading to its being first order in time and space derivatives, the wavefunction ψ has new degrees of freedom and is therefore a 4×1 column vector called a four-component Dirac Spinor:

$$\psi = \begin{pmatrix} \psi_1 \\ \psi_2 \\ \psi_3 \\ \psi_4 \end{pmatrix} \tag{5.82}$$

The conditions on $\vec{\alpha}$ and β ensure that the Hamiltonian is Hermitian:

$$H\Psi = i\hbar \frac{\partial \Psi(r,t)}{\partial t} = c\left(\frac{\hbar}{i}\vec{\alpha}\cdot\vec{\nabla} + \beta mc\right)\Psi(r,t). \tag{5.83}$$

The continuity equation is then given by:

$$\vec{\nabla}\cdot(\psi^\dagger\vec{\alpha}\psi) + \frac{\partial(\psi^\dagger\psi)}{\partial t} = 0 \tag{5.84}$$

in Planck or natural units ($\hbar = 1$ and $c = 1$), where $\psi^\dagger$ is the Hermitian conjugate of ψ given by

$$\psi^\dagger = (\psi_1, \psi_2, \psi_3, \psi_4).$$

Clearly, the probability density ρ and current density $\vec{J}$ are $\rho = \psi^\dagger\psi$ and $\vec{J} = \psi^\dagger\vec{\alpha}\psi$.

Exercise 75

Rewrite the Dirac equation more elegantly by using the four Dirac gamma matrices $\gamma^0 \equiv \beta$; $\gamma^1 \equiv \beta\alpha_x$; $\gamma^2 \equiv \beta\alpha_y$; $\gamma^3 \equiv \beta\alpha_z$; to obtain:

$$(i\gamma^\mu\partial_\mu - m)\psi = 0 \tag{5.85}$$

Hints: 1. Start with the Dirac equation in component form after setting $\hbar = 1$ and $c = 1$ in Planck or natural units:

$$i\frac{\partial \psi(r,t)}{\partial t} = -i\alpha_x \frac{\partial \psi}{\partial x} - i\alpha_y \frac{\partial \psi}{\partial y} - i\alpha_z \frac{\partial \psi}{\partial z} + m\beta \psi. \qquad (5.86)$$

2. Multiply Eq. 5.86 by β. 3. Use the Dirac gamma matrices. 4. Use $\partial_\mu = (\frac{\partial}{\partial t}, \frac{\partial}{\partial x}, \frac{\partial}{\partial y}, \frac{\partial}{\partial z})$.

Klein and Nishina then considered the solution to the Dirac equation for an electron treated as a free particle:

$$\psi(E,p) = v(p)e^{-\frac{1}{\hbar}(Et-(pr))} \qquad (5.87)$$

where E and p denote energy and momentum respectively and satisfy the relativistic energy-momentum relation:

$$m^2 c^2 + p^2 = \frac{E^2}{c^2}; \qquad (5.88)$$

$v(p)$ is a 4x1 matrix that has no explicit space-time dependence and has four solutions, two for positive values of E and two for negative values of E.

Interestingly, the two solutions for each sign of E (positive or negative) indicate degenerate spin states. Klein and Nishina then considered solutions to the Dirac equation for an incident radiation field (beam of photons) using a perturbation method [127]. The incident radiation field is represented by a vector potential $\vec{A}$. The perturbed solution takes the form of a free electron with momentum modified by the incident radiation field. The interaction probability of the field with the electron yields the desired scattering cross-section now widely known as the Klein-Nishina formula.

The Klein-Nishina formula for Compton scattering differential cross-section for Compton scattering is given by:

$$\frac{d\sigma_{KN}}{d\Omega} = \frac{1}{2}r_e^2 \left(\frac{E'}{E}\right)^2 \left[(1+\cos^2 \Theta) + \frac{(E-E')^2}{EE'}\right], \qquad (5.89)$$

where r_e is the classical electron radius. By substituting $2 - sin^2\Theta$ for, $(1 + \cos^2 \Theta)$, we render the Klein-Nishina formula in its more familiar form:

$$\frac{d\sigma_{KN}}{d\Omega} = \frac{1}{2}r_e^2 \left(\frac{E'}{E}\right)^2 \left[\frac{E'}{E} + \frac{E}{E'} - sin^2 \Theta\right] \qquad (5.90)$$

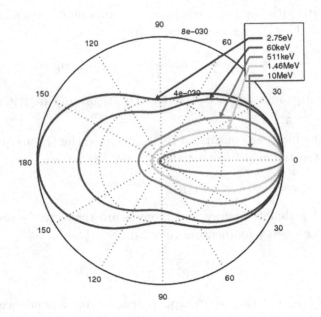

Figure 5.6 Distribution of scattering-angle cross-sections using the Klein–Nishina formula. 2.75 eV visible light rays scatter over all angles; 60 keV X-Rays scatter forward more than backward; 511 keV photons generated via positron annihilation hardly scatter backwards; 1.46 MeV photons from potassium-40 also hardly scatter backward. 10 MeV gamma rays scatter almost entirely forwards. (Free image in public domain, by Dscraggs).

Fig. 5.6) shows the angular distribution of scattering cross-sections using the Klein-Nishina formula. Note that the higher the photon energy, the more an-isotropic the scattering (Fig. 5.6).

5.4 THOMSON AND RAYLEIGH SCATTERING

Both Thomson and Rayleigh scattering are elastic, that is, the energy of the photon does not change, only its direction changes. While Thomson scattering is a photon-electron interaction, Rayleigh scattering is a photon-atom process. Furthermore, both are important at low photon energies where the photoelectric effect is dominant and at mid-energies where Compton scattering dominates. We remind ourselves that for photon energies below 1 MeV the major interaction processes are

Compton scattering, Rayleigh scattering and atomic photoelectric absorption [50]. Like Rayleigh scattering, Thomson scattering is a form of coherent scattering. In essence, Rayleigh scattering is a scattering by the atom as a whole. We now elaborate on Thomson scattering, focusing on mathematical aspects that illuminate the physics. The differential cross-section for Thomson scattering is

$$\boxed{\frac{d\sigma_T}{d\Omega} = \frac{1}{2}r_e^2(1 + \cos^2\Theta)} \tag{5.91}$$

where r_e is the classical electron radius:

$$r_e = \frac{e^2}{4\pi\varepsilon_0 m_e c^2} = 2.818 \times 10^{-15}\text{m}.$$

Note that the Bohr or classical electron radius is obtained by treating the electron as a classical particle while assuming that its rest mass is equivalent to its electrostatic potential. That is,

$$\frac{e^2}{4\pi\varepsilon_0 r_e} = m_e c^2 \tag{5.92}$$

Example 5.18 *Using the Thomson cross-section, calculate the total cross-section, σ_T for scattering into all solid angles, in SI units, in the case of electrons being scattered.*

Solution

Note that the total cross-section is obtained by integrating over all solid angles $d\Omega$.

$$\sigma_T = \int_0^{4\pi} \frac{d\sigma_T}{d\Omega}d\Omega = \frac{2\pi r_e^2}{2}\int_0^{\pi}(1+\cos^2\Theta)d\Theta$$

$$= \frac{8\pi}{3}r_e^2 = 6.65 \times 10^{-29} \text{ m}^2 \tag{5.93}$$

Example 5.19 *Supposing the particle scattered is not an electron but an arbitrary particle of charge q and mass m moving at non-relativistic speed. What is the total Thomson cross-section, σ_T for scattering into all solid angles, for such a particle?*

Solution

$$\sigma_T = \int_0^{4\pi} \frac{d\sigma_T}{d\Omega} d\Omega = \frac{2\pi r_e^2}{2} \int_0^{\pi} (1 + \cos^2\Theta) d\Theta = \frac{8\pi}{3} \left(\frac{q^2}{4\pi\varepsilon_0 mc^2} \right)^2. \quad (5.94)$$

Clearly, we see that the total Thomson scattering cross section is independent of photon energy. The oscillating electric field of the incident photon accelerates the charged particle. Of course, Maxwell's equations [80] predict that electromagnetic waves be generated by oscillating or moving charges. Thus, the charged particle basically absorbs the radiation and emits it at the same frequency as the incident photon or wave, but at a different angle as the incident wave, thereby scattering it.

Exercise 76

Show explicitly that as $E \to E'$, the Klein-Nishina differential cross section formula becomes the Thomson differential cross section.

Exercise 77

Use direct substitution of formulae to express the total Thomson cross section in terms of the Compton wavelength λ_c and the fine structure constant α. You should obtain:

$$\sigma_T = \frac{8\pi}{3} \left(\frac{\alpha\lambda_c}{2\pi} \right)^2 \quad (5.95)$$

We now turn to and focus on Rayleigh scattering initially described by John William Strutt (also known as Lord Rayleigh) [112]. It is also called coherent scattering or elastic scattering since it occurs without loss of energy and therefore without change in frequency or wavelength. It simply depends on atomic structure of the target and the energy of the incident photon. Rayleigh scattering is scattering of photons from harmonically bound electrons, that is, electrons bound to atomic nuclei. It is easily remembered as the reason the sky is blue. This is because blue light is scattered slightly more efficiently than red light by molecules in the sky. The size of the scatterer is much smaller than the wavelength of the incident light. In Rayleigh scattering the oscillating electric field of the incident photon causes polarization of particles, making them to oscillate at the same frequency. The particles therefore become small radiating dipoles giving off these radiations as scattered light. Note that the particles may be individual atoms as well as molecules. Obviously,

Rayleigh scattering is connected with the electric polarizability of materials. Though possible in liquid and solid media, Rayleigh scattering is most prominent in gases. Now, let us get to the mathematical expression which makes the physics tractable. Since Rayleigh scattering involves the atom or molecule as a whole, the charge distribution of all electrons in that atom or molecule must be simultaneously considered in derivations. This is done by introducing the so-called atomic form factor, $F(q,Z)$, to the Thomson differential cross section, leading to a Rayleigh cross-section, σ_R, with the following proportionality:

$$\sigma_R \propto \left(\frac{Z}{\hbar\omega}\right)^2. \tag{5.96}$$

The square of the form factor, $|F(q,Z)|^2$ gives the probality of that the Z electrons of an atom will take up the recoil momentum q without absorbing any energy. Thus, the differential cross-section for Rayleigh scattering of unpolarized photons per solid angle, per atom, is:

$$\boxed{\frac{d\sigma_R}{d\Omega} = \frac{1}{2}r_e^2(1+\cos^2\Theta)|F(q,Z)|^2} \tag{5.97}$$

where r_e is the classical electron radius.

5.5 PAIR PRODUCTION AND BORN APPROXIMATION

Photons with energies E_γ equal to or above $2m_e = 1.022$ MeV can create positron-electron pairs as they interact with matter. The photon is absorbed in the vicinity of the nucleus, and the atom is left in an excited state. Pair production is the dominant photon interaction at high energies (typically above 10 MeV). Pair production takes place in the Coulomb field of a nucleus or electron. For the nuclear field (pair production or nuclear PP), the threshold energy is $2m_e = 1.022$ MeV. However, for the field of the atomic electron (triplet production), the threshold energy is $4m_ec^2 = 2.044$ MeV. The equation for nuclear PP is:

$$T_+ + T_- = E_\gamma - 2m_ec^2 = E_\gamma - 1.02 \text{ MeV}, \tag{5.98}$$

where T_+ and T_- indicate kinetic energies of positron and electron respectively. The equation for electronic PP or triplet production is:

$$T_+ + T_- + T_-(orb) = E_\gamma - 1.02 \text{ MeV}, \tag{5.99}$$

where $T_-(orb)$ is the kinetic energy of the orbital electron in whose field the interaction occurs.

5.5.1 Kinematics of Pair/Triplet Production

The conservation of momentum suggests:

$$\hbar \vec{k}_\gamma = \vec{p}_+ + \vec{p}_-, \tag{5.100}$$

where $\vec{p}_+$ and $\vec{p}_-$ are the momenta of the positron and electron respectively. The conservation of energy suggests:

$$E_\gamma = (T_+ + m_e c^2) + (T_- + m_e c^2), \tag{5.101}$$

where the Ts indicate kinetic energy. Of course, these conditions cannot be satisfied simultaneously, without the atomic nucleus which takes up some momentum as well as some energy in the form of recoil. Note that the existence of the positron was predicted by Dirac's relativistic theory of the electron which allows for negative energy states:

$$E = \pm \left(c^2 p^2 + m_e^2 c^4 \right)^{\frac{1}{2}}. \tag{5.102}$$

The negative states represent a "sea" of electrons which are generally not observable because no transitions into these states can occur due to the Exclusion Principle. When any of the electrons makes a transition to a positive energy level, it leaves a "hole" (positron) which behaves likes an electron but with a positive charge. Of course, the positron is unstable and quickly recombines with an electron in a process called *pair annihilation*, producing two gamma rays in opposite directions (see chapter one on gamma decay, including the importance of pair annihilation in Positron Emission Tomography, PET).

Example 5.20 *A positron and an electron can form an atom called the positronium, with a life time of about 10^{-7} s for parallel spin orientation (ortho-positronium) and about 10^{-9} s for antiparallel orientation (para-positronium). During positron emission tomography, as much as 40% of positron annihilation occurs through the production of positronium atoms inside the patient's body. Since the decay of these positronium atoms is sensitive to metabolism, there is ongoing research aimed at taking advantage of positronium decay for diagnosis and prognosis of diseases [82]. The positronium is also an ideal system in the quest for new physics beyond the standard model and especially for research into bound-state quantum electrodynamics (QED) [83]. State in outline some* *quantum mechanical differences between Positronium (Ps) and Hydrogen (H) in order to illuminate the particle-antiparticle nature of Ps and give some theoretical basis for Ps Imaging [111].*

Solution

1. The reduced mass μ_{Ps} of the positronium is

$$\mu_{Ps} = \frac{(m_{e-})(m_{e+})}{m_{e-} + m_{e+}} = \frac{m_e^2}{2m_e} = \frac{m_e}{2}$$

   since the mass of electron and positron are identical but the reduced mass for H stays as

$$\mu_H = \frac{m_e m_p}{m_e + m_p},$$

   where m_p is the mass of the proton.

2. Due to this reduced mass ratio, the Schrodinger energy levels, are

$$E_n(Ps) \approx \frac{1}{2} E_n(H),$$

   hence,

$$E_n(Ps) = \frac{m_e c^2 \alpha^2}{4n^2}$$

   where α is the fine structure constant.

3. The binding energy of Ps is about 6.8 eV compared to 13.6 eV for H, hence

$$E_n(Ps) = \frac{-6.8 \ eV}{n^2},$$

   while

$$E_n(H) = \frac{-13.6 \ eV}{n^2}.$$

Example 5.21 *Using considerations of energy and momentum conservation as in Eq 5.103 and Eq. 5.106 show that pair production cannot occur in free space.*

Solution

Conservation of momentum and energy require:

$$p_i = p_f,$$

$$E_i = E_f,$$

for initial (i) and final (f) quantities. The conservation of energy suggests:

$$E_\gamma = (T_+ + m_e c^2) + (T_- + m_e c^2). \tag{5.103}$$

For the simple case of identical kinetic energies for positron and electron (in general, they could be different):

$$E_\gamma = (T_+ + m_e c^2) + (T_- + m_e c^2) = 2\gamma m_e c^2. \tag{5.104}$$

Applying these to photon momentum/energy and those of the produced positron and electron, we get, for this simple case of identical kinetic energies of produced positron and electron:

$$\hbar \vec{k_\gamma} = \frac{E_\gamma}{c} \equiv \frac{E_{PP}}{c} = \frac{2\gamma m_e c^2}{c} = \left(\frac{2\gamma m_e v}{v}\right) = (p_+ + p_-)\frac{c}{v} \tag{5.105}$$

which violates conservation of momentum because v is always less than c, so that photon momentum ends up larger than the combined momentum of positron-electron pair. That is, Eq. 5.105 violates the following expectation for momentum conservation:

$$\hbar \vec{k_\gamma} = \vec{p_+} + \vec{p_-}, \tag{5.106}$$

where $\vec{p_+}$ and $\vec{p_-}$ are the momenta of the positron and electron respectively. Note that the same violation is obtained in the general case of different kinetic energies of the positron-electron pair.

5.5.2 Differential Cross Section for Pair Production

The differential cross section for the production of an electron-positron pair in the field of an unscreened Coulomb potential energy $U = \frac{-ke^2}{r}$, was first calculated using first Born approximation in early papers by Heitler and Sauter [48, 104]. Drawing analogy from the theory of Bremsstrahlung for the transition probability where the incident particle is a photon instead of an electron as in Bremsstrahlung, and using appropriate density of states for the emission of the positron-electron pair based on

Dirac's relativistic quantum theory for the electron and Born approximation, Heitler and Sauter obtained the following energy differential cross-section for the positron in pair production:

$$\frac{d\sigma_{pp}}{dT_+} = \frac{4}{137} r_e^2 Z^2 \left(\frac{T_+^2 + T_-^2 - \frac{2}{3} T_+ T_-}{E_\gamma^3} \right)^2 \left[\frac{E'}{E} + \frac{E}{E'} - \sin^2 \Theta \right] \quad (5.107)$$

In their 1933 letter to Nature [48] with the very instructive title *Stopping of Fast Particles with Emission of Radiation and the Birth of Positive Electrons*, Heitler and Sauter integrated Eq. 5.107 to obtain the total pair production cross section in the nuclear field:

$$\sigma_{pp} = \int \left(\frac{d\sigma_{pp}}{dT_+} \right) dT_+ = \frac{Z^2}{137} r_e^2 \left[\frac{28}{9} \ln \frac{2E_\gamma}{mc^2} - \frac{218}{27} \right]$$

$$= \sigma_0 Z^2 \left[\frac{28}{9} \ln \frac{2E_\gamma}{mc^2} - \frac{218}{27} \right] \quad (5.108)$$

where $\sigma_0 = \frac{r_e^2}{137}$. Eq. 5.108 is valid when $mc^2 << E_\gamma << 137mc^2/\sqrt{Z}$. It does not consider screening effects in the nuclear field, that is, the partial reduction of the nuclear charge by the electric potential of the inner-shell electrons. When screening is considered, a reduction in cross section is expected and for $E_\gamma >> 137mc^2/\sqrt{Z}$ the result is:

$$\sigma_{pp} = \sigma_0 Z^2 \left[\frac{28}{9} \ln \frac{183}{Z^{1/3}} - \frac{218}{27} \right] \quad (5.109)$$

Current accepted data for pair and triplet production cross sections of attenuation coefficients of photons with various elements, compounds and mixtures (absorbers) are available at NIST XCOM which include data from various authors such as Hubbard et al., 1980, (see ICRU Report 90).

Exercise 78

Computed Tomography (CT) has had tremendous impact in medical diagnostics and other fields since its invention in the 1970s by Godfrey Hounsfield and Allan Cormack (Nobel laureates in Physiology or Medicine in 1979). Mathematically, CT acquisition is the collection of the line integrals, p, of the 3D distribution of an object's attenuation coefficient, $\mu(x,y,z,E,t)$, through all the possible lines l crossing the object [88]:

$$p_l(E,t) = \int_l \mu(x,y,z,t,E)dl. \qquad (5.110)$$

where E is the X-ray photon energy and t is time. Assume an ideal mono-energetic distribution of the photons in the incident beam, use Beer's law of exponential attenuation to find an expression for $p_l(E,t)$ in terms of the measured X-ray intensities of the un-attenuated beam, I_0 the attenuated beam $I(t)$ recorded along the line l at time t.

Exercise 79

Recently, 3D breast imaging (tomosynthesis and cone beam breast CT) has been introduced in clinical practice with significant improvement in diagnostic accuracy and there are ongoing studies on the use of phase-contrast techniques to improve the image quality of breast CT [71, 72, 79]. An important example of phase-contrast techniques is propagation based imaging, PBI, which is based on propagation of the wave vector in free space. (a) What physical phenomena treated in this chapter are avoided by PBI? (b) Describe the variables in the following expression for the complex refractive index n which captures phase changes and attenuation in the material:

$$n(r,E) = 1 - \delta(r,E) + i\beta(r,E) \qquad (5.111)$$

Exercise 80

X-Ray PCI (phase contrast imaging) or XPCI takes advantage of the wave nature of X-rays and is generated due to the phase shift of an incident wave when passing through an object. This phase shift is described by the real part of the complex index of refraction [42, 101] as shown in Eq. 5.111. In line with the wave function formalism shown in this chapter, the propagation of an electromagnetic wave with amplitude E_0 and wave vector k through a medium with refractive index $n = 1 - \delta + i\beta$ can be described thus:

$$\Psi(r) = E_0 e^{inkr} \qquad (5.112)$$

Expand the wave function to obtain separate exponential terms for phase shifts and attenuation, which is what is done in the image analysis to obtain diagnostic information.

5.6 ANSWERS

Answer of exercise 66

Solution $0.693/5cm = 0.1386/cm$.

Answer of exercise 67

Steps given in the text.

Answer of exercise 68

Solutions (a) $E_B(K) = 69.5$ keV, $E_B(L1) = 12.1$ keV, $E_B(L2) = 11.54$ keV, and $E_B(L3) = 10.21$ keV. (b) $P(K) = 0.804$ or 80.4%, $P(L1) = 0.026$ or 2.6%, $P(L2) = 0.047$ or 4.7%. $P(L) = 0.076$ and therefor $P(L) = 0.149$ or 14.9%. (c) When incident energy is less than $E_B(K)$ but above or equal to $E_B(L1)$, then the 2 K-shell electrons will not participate in photoelectric effect, hence, $P(K) = 0$ and the probabilities are still calculated as in (b) but with $P(K) = 0$ in the summation.

Answer of exercise 69

Ans: 0.486 MeV

Answer of exercise 70

Ans: (a) $KE_{electron} = 0.0275$ MeV. (b) 0.1 MeV $- 0.0275$ MeV $= 0.0725$ MeV.

Answer of exercise 71

Steps given in the text.

Answer of exercise 72

Steps given in the text.

Answer of exercise 73

Steps given in the text.

Answer of exercise 74

Ans: 0.275 MeV

Answer of exercise 75

Steps given in the text.

Answer of exercise 76

Solution is straightforward.

Answer of exercise 77

Steps given in the text.

Answer of exercise 78

Solution $p_l(E,t) = -\ln(I(t)/I_0)$

Answer of exercise 79

(a) Attenuation and scattering, since, empty air or vacuum environment is introduced between sample and detector (b) In Eq. 5.111, r and E denote positions inside the object and the photon energy; δ denotes the real decrement responsible for the phase shift, i is the imaginary unit such that β denotes the imaginary part which characterizes the photon attenuation in the material.

Answer of exercise 80

Solution

$$\Psi(r) = E_0 e^{inkr} = E_0 e^{i(1-\delta)kr} e^{-\beta kr} \tag{5.113}$$

General Mathematical Definitions and Derivations

A.1 CROSS SECTIONS

A.1.1 Scattering and Absorption Cross Sections

In nuclear, atomic, and particle physics, the differential and total scattering cross sections are among the most important measurable quantities. ICRU 85 [116] gives a general definition for the cross-section, σ of a target entity, for a particular interaction (eg, Compton scattering, Rutherford scattering) produced by incident charged or uncharged particles of a given type and energy as the quotient of N by Φ:

$$\sigma = \frac{N}{\Phi},\tag{A.1}$$

where N is the mean number of such interactions per target entity subjected to the particle *fluence*, Φ which is the number of particles incident on a sphere of cross-sectional area da. Note that the fluence Φ is the time integrated flux, ϕ (number of particles per second, per area), so every definition involving fluence can be replaced with flux along with appropriate rates. The fluence itself is given by:

$$\Phi = \frac{dN}{da}\tag{A.2}$$

where dN is the number of particles incident on a sphere of cross-sectional area da and is usually a function of the angles θ and ϕ in spherical

DOI: 10.1201/9781003215622-A

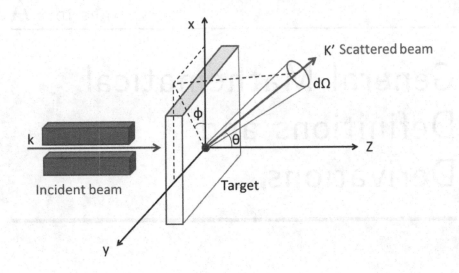

Figure A.1 Schematic for measurement and calculation of differential scattering cross-section. An incident beam scatters from the target material into a solid angle element $d\Omega$.

coordinates (see Fig. A.1). The unit of σ is m^2. A special unit commonly used for convenience is the barn, b, with

$$1\,\mathrm{b} = 10^{-28}\,\mathrm{m}^2 = 100\,\mathrm{fm}^2.$$

For completeness, the fluence Φ in σ relates to the flux ϕ by considering a change in fluence over an infinitesimally small interval such that

$$\phi(t) = \frac{d\Phi}{dt} = \frac{d}{dt}\left(\frac{dN}{da}\right), \tag{A.3}$$

and the fluence becomes:

$$\Phi(t_0, t_1) = \int_{t_0}^{t_1} \phi(t)\,dt. \tag{A.4}$$

For a scattering process, the cross section becomes the *scattering cross section*. For an absorption process, it is termed the *absorption cross section*. The cross section can even characterize a process of transformation to another species. Physically, the cross section is a measure of the probability that a specific process will take place in a collision or interaction of two entities usually particles.

A.1.2 Differential Cross Sections

The differential cross section $d\sigma/d\Omega$ is the ratio of the number of scattered particles $dN(\theta,\phi)$ per unit time within the solid angle $d\Omega$ divided by the incident particle fluence, Φ:

$$\frac{d\sigma}{d\Omega} = \frac{dN(\theta,\phi)}{\Phi d\Omega}, \tag{A.5}$$

The differential cross section can also be expressed in terms of the energy and momentum of particles following interaction. Thus, differential cross section is a way of specifying the cross section as a function of convenient final-state variables, such as particle angle or energy. When a cross section is integrated over all scattering angles (and possibly other variables), it is called a total cross section.

A.1.3 Total Cross Sections

The total cross section σ_T is obtained by integrating the differential cross section over all scattering angles (and any other variable of interest):

$$\sigma_T = \oint_{4\pi} \frac{d\sigma}{d\Omega} d\Omega = \int_0^{2\pi} \int_0^{\pi} \frac{d\sigma}{d\Omega} \sin\theta d\theta d\phi. \tag{A.6}$$

A better term for σ_T in Eq. A.6 might be *integral cross section* but conventional use of *total cross section* is ubiquitous and allow context to distinguish σ_T from the case where incident particles of a given type and energy undergo different and independent interaction types in a target material, leading to the sum of component cross sections σ_j, which we denote again as σ, given by:

$$\sigma = \sum_j \sigma_j = \frac{1}{\Phi}\sum_j N_j, \tag{A.7}$$

where N_j is the average number of type J interactions per target entity subjected to the particle fluence Φ and σ_j is the component cross section for type j interaction.

A.2 SCHRÖDINGER'S EQUATION AND CROSS SECTIONS

Consider the time dependent Schrödinger equation for a non-relativistic particle in position basis:

$$i\hbar \frac{\partial \Psi(\mathbf{r},t)}{\partial t} = \left[-\frac{\hbar^2}{2m}\nabla^2 + V(r,t) \right] \Psi(r,t) \tag{A.8}$$

for a specified potential energy function (often just called *potential* which can be confusing since the same symbol and term often refer to electric potential or electric potential energy per charge), V(r,t), with m the mass of particle and ∇^2 the Laplacian. Eq. A.8 has the form of a diffusion equation, with an imaginary constant present in the transient term. Using Born's statistical interpretation of quantum mechanics, the complex conjugate of the wave function gives us the probability of finding the particle at position **r** at time t:

$$\Psi(\mathbf{r},t)^*\Psi(\mathbf{r},t) = |\Psi(\mathbf{r},t)|^2 \tag{A.9}$$

or more precisely,

$$\int_a^b \Psi(\mathbf{r},t)^*\Psi(\mathbf{r},t)\mathrm{d}x = \int_a^b |\Psi(\mathbf{r},t)|^2\mathrm{d}x \tag{A.10}$$

gives the probability of finding the particle between positions a and b at time t. Of course, the particle is somewhere so normalization ensures that

$$\int_{-\infty}^{+\infty} |\Psi(\mathbf{r},t)|^2\mathrm{d}x = 1. \tag{A.11}$$

To enable separation of variables, the complex-valued wave function can be expressed as:

$$\Psi(\mathbf{r},t) = \psi(r)\phi(t) = \psi(r)e^{-i\omega t} = \psi e^{-i\frac{E}{\hbar}t}. \tag{A.12}$$

where, in our case, the particle has kinetic energy E outside the range of the potential energy:

$$E = \frac{1}{2}mv^2 = \frac{\hbar^2 k^2}{2m} = \hbar\omega \tag{A.13}$$

and $\hbar k = mv$. Based on Eq. A.12, we focus on the time-independent Schrödinger equation:

$$-\frac{\hbar^2}{2m}\nabla^2\psi + V(r)\psi = E\psi \tag{A.14}$$

which can be rewritten as:

$$\nabla^2\psi + \frac{2m}{\hbar^2}(E - V(r))\psi = 0. \tag{A.15}$$

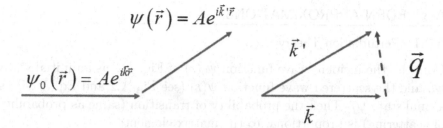

Figure A.2 First order Born approximation where a plane wave $\psi_0(r)$ is scattered into a wave of same amplitude, $\psi(r)$ as the incoming wave but with different direction.

Eq. A.15 is a mathematical special case of the Helmholtz's equation with proper choice of k^2;

$$\nabla^2 y + k^2 y = 0 \qquad (A.16)$$

and we have many ways of solving Eq. A.16 depending on the physical system. For *elastic* scattering problems, we start by searching for solutions for values of r large compared to the range of the potential energy $V(r)$:

$$\psi(r) = e^{ikz} + \frac{f(\theta,\phi)}{r} e^{ikr}. \qquad (A.17)$$

If the terms of the right hand side of Eq. A.17 are each multiplied by $e^{-i\omega t}$, to get back the full wave function, Eq. A.12, then the first term of Eq. A.17 represents the incident particle coming in along the z-axis (Fig. A.2) and the second term is a spherical wave traveling outwards from the origin or target and thus represents the scattered particle. The condition for having a solution in the form of Eq. A.17 includes having $V(r)$ go to zero faster than $\frac{1}{r}$, as $r \xrightarrow{\infty}$. In Eq. A.17, $f(\theta,\phi)$ is the *scattering amplitude* and it has an exciting relationship with the scattering cross-section:

$$\frac{d\sigma}{d\Omega} = |f(\theta,\phi)|^2. \qquad (A.18)$$

Measurements can provide the left hand side of Eq. A.18 from detectors (see Fig. A.1) and theory can yield the right hand side, for comparisons. How to calculate the scattering amplitude so as to make use of Eq. A.18 leads to the need for Born approximation.

A.3 BORN APPROXIMATIONS

A.3.1 Perturbation Theory

Consider the incident wave function $\psi_0(r)$ of Fig. A.2 as an initial state, ψ_i, and the scattered wave function $\psi(r)$ (see Fig A.2 and Eq. A.34) as a final state ψ_f. Then the probability of transition (same as probability of scattering) is proportional to the matrix element:

$$\langle \psi_f | V(\mathbf{r},t) | \psi_i \rangle \propto \int_0^\infty d^3r V(r) e^{iq} \qquad (A.19)$$

In Chapter 1, subsection 1.6.3, we used time-independent perturbation theory in outline to obtain the so-called Fermi's 2nd golden rule for the transition rate or probability of transition between states for a given perturbation. The matrix elements for the probability of scattering in Eq. A.19 are equivalent to the matrix elements we defined for the perturbed part of the Hamiltonian, $H'_{ba} \equiv \langle \phi_b | H' | \phi_a \rangle$ subsection 1.6.3. Thus, as in subsection 1.6.3 Fermi's 2nd golden rule gives the transition rate $\Gamma_{i \to f}$ from an initial state ψ_i to a final state ψ_f. For a system that obey's the time-dependent Schrödinger's equation (Eq. A.8), time-dependent perturbation theory holds that for different initial and final states, denoted by subscripts i and f respectively,

$$\langle \psi_f | \psi \rangle = \delta_{fi} - \frac{i}{\hbar} \int_0^t \langle \phi_f | H'(t') | \phi_f \rangle \, e^{i\omega_{fi}t'} \, dt', \qquad (A.20)$$

where the Hamiltonian $H'(t) = H_0 + V(r,t)$ and $V(r,t)$ is the perturbation. For weak perturbations and long times, $H'(t) = H'$ (constant in time), the transition probability $P_f(t)$ is given by

$$\Gamma_{i \to f} \times t, \; for \; f \neq i. \qquad (A.21)$$

To obtain an expression for $\Gamma_{i \to f}$ in this scattering/absorption case we redefine the perturbation in subsection 1.6.3, $H'_{ba} \equiv \langle \phi_b | H' | \phi_a \rangle$ as $H'_{fi} \equiv \langle \psi_f | H' | \psi_i \rangle = \langle \psi_f | V(r) | \psi_i \rangle$, since $V(r)$ is only present during interaction and absent from the unperturbed Hamiltonian. Thus, the transition rate, obtained as probability per time, for going from i to any other state f is again (as in subsection 1.6.3) a constant given by Fermi's 2nd golden rule:

$$\boxed{\Gamma_{i \to f} = \frac{2\pi}{\hbar} | \langle f | V | i \rangle |^2 \int_0^\infty \rho(E_f) \delta(E_f - E_i) dE_f,} \qquad (A.22)$$

where the matrix element $|\langle f|V|i\rangle|$ is approximately constant within the energy range $|(E_f - E_i)| \leq 2\pi\hbar/t$ of the allowed final states f.

We can now use Fermi's golden rule to derive the differential scattering cross section for a particle of mass m with incident wavefunction ψ_i and wave vector $\vec{k}$ scattering from a potential $V(r)$ into a solid angle (Fig. A.1) where the scattering can be observed by a detector. Let the initial wavefunction be "box-normalized" plane wave in the limit that the dimension of the box $a \to \infty$:

$$\psi_i = \frac{1}{\sqrt{a^3}} e^{i\vec{k}.\vec{r}}, \tag{A.23}$$

and likewise the final wavefunction:

$$\psi_f = \frac{1}{\sqrt{a^3}} e^{i\vec{k}'.\vec{r}}. \tag{A.24}$$

Thus, the matrix elements of our perturbation,

$$\langle \psi_f|H'|\psi_i\rangle = \langle \psi_f|V(r)|\psi_i\rangle$$

become:

$$\langle \psi_f|V(r)|\psi_i\rangle = \int \psi_f^* V(\vec{r}) \psi_i d^3\vec{r} = \frac{1}{a^3} \int e^{i\vec{q}.\vec{r}} V(\vec{r}) d^3\vec{r}, \tag{A.25}$$

where

$$\vec{q} = \vec{p}' - \vec{p} = \vec{k}' - \vec{k}.$$

Next we calculate the density of states, $\rho(E_f))dE_f$, in Eq. A.22, that is, number of states with energies between E and $E + dE_f$ with wave vector $\vec{k}'$ within the detector, that is, within $d\Omega$ such that the volume is

$$k'^2 dk' d\Omega.$$

Since each state in k-space occupies a volume $(\frac{2\pi}{a})^3$, then the number of states in $k'^2 dk' d\Omega$ is

$$\rho(E_f))dE_f = \frac{k'^2 dk' d\Omega}{(\frac{2\pi}{a})^3}. \tag{A.26}$$

Dividing Eq. A.26 by dE and using

$$E = \frac{\hbar^2 k^2}{2m}$$

for a free particle, we have:

$$\rho(E_f) = \frac{a^3}{(2\pi)^3} \frac{\sqrt{2m^3E}}{\hbar^3} d\Omega. \tag{A.27}$$

Hence, applying Fermi's 2nd golden rule (Eq A.22), with substitutions (Eq. A.25 and Eq. A.27) the rate at which particles scatter into the detector with solid angle $d\Omega$ is:

$$\Gamma_{i \to d\Omega} = \frac{2\pi}{\hbar} |\langle f|V|i \rangle|^2 \int_0^\infty \rho(E_f) \delta(E_f - E_i) dE_f$$

$$= \frac{2\pi}{\hbar} \left| \frac{1}{a^3} \int e^{i\vec{q}\cdot\vec{r}} V(\vec{r}) d^3\vec{r} \right|^2 \frac{a^3}{(2\pi)^3} \frac{\sqrt{2m^3E}}{\hbar^3} d\Omega. \tag{A.28}$$

Now, from Eq. A.23, the incident plane wave is of the form:

$$\psi_i = \frac{1}{\sqrt{a^3}} e^{i\vec{k}\cdot\vec{r}} = A e^{i\vec{k}\cdot\vec{r}} \tag{A.29}$$

and so we apply the definition of probability current for incident wave J_i thus:

$$J_i = |A|^2 \frac{\hbar \vec{k}}{m} = \frac{1}{a^3} \frac{\hbar \vec{k}}{m}. \tag{A.30}$$

It is now easy to see that the differential scattering cross section is related to the ratio of the rate of scattering into $d\Omega$ and the incident probability current thus:

$$\frac{d\sigma}{d\Omega} = |f(\theta,\phi)|^2 = \frac{\Gamma_{i \to d\Omega}}{J_i d\Omega}. \tag{A.31}$$

Substituting Eq. A.28 and Eq. A.30 into Eq. A.31, we obtain a recognizable result for the scattering cross section using Fermi's 2nd golden rule from perturbation theory:

$$\boxed{\frac{d\sigma}{d\Omega} = |f(\theta,\phi)|^2 = \left| \frac{m}{2\pi\hbar^2} \int e^{i\vec{q}\cdot\vec{r}} V(\vec{r}) d^3\vec{r} \right|^2.} \tag{A.32}$$

A.3.2   First Born Approximation for Scattering Amplitude

Born approximation is ubiquitous for scattering problems in atomic, nuclear particle physics [53, 63] and beyond. Here, we illustrate and state in general terms the Born approximation for the scattering factor f for a potential $V(r)$. As shown in Fig.A.2, the incoming wave is considered a plane wave of direction $\vec{k}$ which upon interaction gets deflected in a direction $\vec{k}'$ but keeps same amplitude and same wavelength. The scattering vector $\vec{q}$ is given by $\vec{k}' - \vec{k}$ and its magnitude is

$$q = 2k\sin\left(\frac{\theta}{2}\right) = \frac{4\pi}{\lambda}\sin\left(\frac{\theta}{2}\right). \tag{A.33}$$

Eq. A.33 was obtained by explicitly accounting for the change in momentum, ie, final minus initial momentum, $\vec{p}' - \vec{p}$ as illustrated in Fig.A.2, thus:

$$|\vec{q}| = |\vec{p}' - \vec{p}| = |\vec{k}' - \vec{k}| = \sqrt{(k')^2 + k^2 - 2k'k\cos\theta}$$
$$= k\sqrt{2(1 - \cos\theta)} = 2k\sin\left(\frac{\theta}{2}\right). \tag{A.34}$$

Using $\vec{q}$ as defined and with the physical meaning of momentum transfer, the scattering factor for the *First Born approximation* is given by

$$\boxed{f_{Born}(\theta, \phi) = \frac{-2m}{\hbar^2}\frac{1}{4\pi}\int e^{i\vec{q}\cdot\vec{r}}V(\vec{r})d^3\vec{r}.} \tag{A.35}$$

Note that the scattering factor or scattering amplitude f_{Born} is proportional to the Fourier transform of the potential $V(r)$ with respect to q [90]. Both the physics and the mathematics are beautiful: as the scattering amplitude modifies the outgoing spherical wave, it engenders the physical observable of the differential cross section which the detector records.

A.3.3   Born Series using Green's Function

Bringing back Eq. A.15, as a mathematical special case of the Helmholtz's equation with proper choice of k^2; and rewriting it to show the homogeneous equation with incident k^2, we have:

$$(\nabla^2 + k^2)\psi_0 = 0 \quad where \; k^2 = \frac{2mE}{\hbar^2} \tag{A.36}$$

and the incident plane wavefunction is

$$\psi_0 = Ae^{i\vec{k}\cdot\vec{r}}. \tag{A.37}$$

The in-homogeneous equation accounts for the interaction or scattering such that

$$(\nabla^2 + k^2)\psi(\vec{r}) = Q \quad where \ k^2 = \frac{2mE}{\hbar^2}, \quad Q = \frac{2m}{\hbar^2}V(\vec{r})\psi(\vec{r}). \quad (A.38)$$

Thus, the general solution, $\psi(r)$ for Eq. A.15 can be presented as the sum of two components, namely, (i) solution to the homogeneous equation, (Eq. A.36) which is Eq. A.37, and (ii) a particular solution with the interaction potential $V(r)$, using Green's function:

$$\psi(\vec{r}) = \psi_0(\vec{r}) + \frac{2m}{\hbar^2}\int G(\vec{r}-\vec{r}')V(\vec{r}')\psi(\vec{r}')d^3r'. \quad (A.39)$$

Of course, the Green's function $G(\vec{r})$ in general solves Helmholtz equation (Eq. A.38) with a point source or delta function source:

$$(\nabla^2 + k^2)G(\vec{r}) = \delta^3(\vec{r}). \quad (A.40)$$

Here is the proof. Substituting $Q(\vec{r})$ into the particular solution part of Eq. A.39 based on our definition in Eq. A.38, we have,

$$\psi(\vec{r}) = \int G(\vec{r}-\vec{r}')Q(\vec{r}')d^3r', \quad (A.41)$$

which enables us to see that,

$$(\nabla^2 + k^2)\psi(\vec{r}) = \int [(\nabla^2 + k^2)G(\vec{r}-\vec{r}')]Q(\vec{r}')d^3r'$$
$$= \int \delta^3(\vec{r}-\vec{r}')Q(\vec{r}')d^3r' = Q(\vec{r}). \quad (A.42)$$

We now need the explicit form of the Green's function that solves Eq. A.40. There are many ways of obtaining it such as using Fourier transformation to turn the differential equation into an algebraic one and then carrying out contour integration (method of residues), including the use of Cauchy's integral formula. Many standard Quantum Mechanics texts such as Griffiths [43] and Shankar [110] give this procedure explicitly and the curious reader is referred to these. We proceed with the results, namely, the two explicit forms of Green's function we need:

$$G_+(\vec{r}-\vec{r}') = -\frac{1}{4\pi}\frac{e^{ik|\vec{r}-\vec{r}'|}}{|\vec{r}-\vec{r}'|}, \quad (A.43)$$

representing an outgoing spherical wave emitted from $\vec{r}'$, and

$$G_-(\vec{r} - \vec{r}') = -\frac{1}{4\pi} \frac{e^{-ik|\vec{r} - \vec{r}'|}}{|\vec{r} - \vec{r}'|}, \tag{A.44}$$

representing an incident wave that converges onto $\vec{r}$. In our scattering case, we have only outgoing scattered waves and so we use $G_+(\vec{r} - \vec{r}')$ in the general solution of Eq. A.39 to get:

$$\boxed{\psi(\vec{r}) = \psi_0(\vec{r}) - \frac{m}{2\pi\hbar^2} \int \frac{e^{ik|\vec{r} - \vec{r}'|}}{|\vec{r} - \vec{r}'|} V(\vec{r}') \psi(\vec{r}') d^3 r'.} \tag{A.45}$$

Eq. A.45 is just an integral form of the Schrödinger's equation (Eq. A.14) which can be solved by means of successive or iterative approximations aptly called *Born series*.

A.3.3.1 Zeroth Order Solution

$$\psi(\vec{r})_0 = \psi_0(\vec{r}). \tag{A.46}$$

A.3.3.2 First Order Solution

$$\psi(\vec{r})_1 = \psi_0(\vec{r}) - \frac{m}{2\pi\hbar^2} \int \frac{e^{ik|\vec{r} - \vec{r}_1|}}{|\vec{r} - \vec{r}_1|} V(\vec{r}_1) \psi_0(\vec{r}_1) d^3 r_1, \tag{A.47}$$

obtained by inserting $\psi(\vec{r})_0 = \psi_0(\vec{r})$ into Eq. A.45. This is the *first Born approximation* where an exact part of the solution, the scattered wavefunction, $\psi(\vec{r})_0$, is replaced by the known incident wave function $\psi_0(\vec{r})$.

A.3.3.3 Second Order Solution

This is obtained by inserting the first order solution $\psi(\vec{r})_1$ into Eq. A.45:

$$\begin{aligned}
\psi(\vec{r})_2 &= \psi_0(\vec{r}) - \frac{m}{2\pi\hbar^2} \int \frac{e^{ik|\vec{r} - \vec{r}_2|}}{|\vec{r} - \vec{r}_2|} V(\vec{r}_2) \psi(\vec{r}_2)_1 d^3 r_2 \\
&= \psi_0(\vec{r}) - \frac{m}{2\pi\hbar^2} \int \frac{e^{ik|\vec{r} - \vec{r}_2|}}{|\vec{r} - \vec{r}_2|} V(\vec{r}_2) \psi_0(\vec{r}_2) d^3 r_2 \\
&\quad + \left(\frac{m}{2\pi\hbar^2}\right)^2 \int \frac{e^{ik|\vec{r} - \vec{r}_2|}}{|\vec{r} - \vec{r}_2|} V(\vec{r}_2) \psi(\vec{r}_2)_1 d^3 r_2 \\
&\quad \times \int \frac{e^{ik|\vec{r}_2 - \vec{r}_1|}}{|\vec{r}_2 - \vec{r}_1|} V(\vec{r}_1) \psi_0(\vec{r}_1) d^3 r_1.
\end{aligned} \tag{A.48}$$

A.3.3.4 Nth Order Solution

Similar iterations give us the Nth order Born approximation for the wave function. To make this transparent, we make the following substitution. Let

$$g \equiv -\frac{m}{2\pi\hbar^2} \frac{e^{ik|\vec{r}-\vec{r}'|}}{|\vec{r}-\vec{r}'|}. \tag{A.49}$$

Then Eq. A.45 can be written schematically thus:

$$\psi = \psi_0 + \int gV\psi. \tag{A.50}$$

Now taking Eq. A.50 as an expression for ψ and using it to replace the last term of the same Eq. A.50, we get:

$$\psi = \psi_0 + \int gV\psi_0 + \int\int gVgV\psi. \tag{A.51}$$

The interation of this procedure gives the Born series for ψ to any desired *nth order* in which only the incident wavefunction ψ_0 appears in each integrand, along with increasing powers of gV:

$$\boxed{\psi = \psi_0 + \int gV\psi_0 + \int\int gVgV\psi_0 + \int\int\int gVgVgV\psi_0 + ...} \tag{A.52}$$

The Born series captures the physical situation where an incoming particle undergoes a sequence of multiple scattering events from the potential. In practice, one scattering is considered and that corresponds to the first order solution or *first Born approximation*. Clearly, the *First Born Approximation* given in subsection A.3.2, and from the Born's series in this subsection A.3.3 as well as that from Fermi's 2nd golden rule using perturbation theory (subsection A.3.1) all coincide.

Experimental Data: From Creighton University NIM Lab

B.1 DATA: CHAPTER 2

B.1.1 Data for Fig. 2.6

DOI: 10.1201/9781003215622-B

A. Dwell time: 0.02 s

Cell #	Count Range	Frequency	Tot Obs.	Dev. Sq.	Poisson Pred.	Resid.	X^2	Gauss Pred.	Resid.	X^2
1	0 – 0	68	0	74.23	70.34	-2.34	0.078	46.29	21.71	10.19
2	1 – 1	76	76	0.15	73.50	2.50	0.085	77.98	-1.98	0.05
3	2 – 2	38	76	34.66	38.41	-0.41	0.004	50.45	-12.45	3.07
4	3 – 3	15	45	57.33	13.38	1.62	0.197	12.54	2.46	0.48
4	4 – 4	3	12	26.20	3.50	-0.50	0.070	1.20	1.80	2.72

B. Dwell time: 0.04 s

Cell #	Count Range	Frequency	Tot Obs.	Dev. Sq.	Poisson Pred.	Resid.	X^2	Gauss Pred.	Resid.	X^2
1	0 – 0	22	0	109.40	21.51	0.49	0.01	17.52	4.48	1.15
2	1 – 1	47	47	71.11	47.96	-0.96	0.02	38.06	8.94	2.10
3	2 – 2	53	106	2.80	53.47	-0.47	0.00	52.80	0.20	0.00
4	3 – 3	40	120	23.72	39.75	0.25	0.00	46.78	-6.78	0.98
5	4 – 4	23	92	72.06	22.16	0.84	0.03	26.47	-3.47	0.45
6	5 – 5	10	50	76.73	9.88	0.12	0.00	9.564	0.44	0.02
7	6 – 6	4	24	56.85	3.67	0.33	0.03	2.21	1.79	1.46
7	7 – 7	1	7	22.75	1.17	-0.17	0.03	0.33	0.68	1.40

Figure B.1 Sample experimental results for runs with dwell times 0.02 s and 0.04 s. The results were binned into cells with specified count range and their frequencies. The Poisson and Gaussian predicted values were calculated using Eq. 2.33 and Eq. 2.44 respectively. The Chi-Squared values were obtained using Eq. 2.82 based on the residuals shown in the table. Independent cell numbers were assigned only for cells containing at least 5 measurements and those below 5 were assigned the next lower cell number having 5 or more measurements. Data was taken using a NaI(Tl) scintillation detector with multi-channel analyzer set for multi-scaling, by Creighton University NIM students, Fall 2019, analyzed and tabulated by Zachary J. Sabata.

Nuclear Physics Databases: E-sources

C.1 ATOMIC WEIGHTS AND ISOTOPIC COMPOSITIONS WITH RELATIVE ATOMIC MASSES

Atomic Weights and Isotopic Compositions Developers and Contributors: J. S. Coursey, D. J. Schwab, J. J. Tsai, and R. A. Dragoset, NIST Physical Measurement Laboratory.

DOI: 10.1201/9781003215622-C

Atomic Weights and Isotopic Compositions for All Elements

Isotope		Relative Atomic Mass	Isotopic Composition	Standard Atomic Weight	Notes
1 H	1	1.007 825 032 23(9)	0.999 885(70)	[1.007 84, 1.008 11]	m
D	2	2.014 101 778 12(12)	0.000 115(70)		
T	3	3.016 049 2779(24)			
2 He	3	3.016 029 3201(25)	0.000 001 34(3)	4.002 602(2)	g,r
	4	4.002 603 254 13(6)	0.999 998 66(3)		
3 Li	6	6.015 122 8874(16)	0.0759(4)	[6.938, 6.997]	m
	7	7.016 003 4366(45)	0.9241(4)		
4 Be	9	9.012 183 065(82)	1	9.012 1831(5)	
5 B	10	10.012 936 95(41)	0.199(7)	[10.806, 10.821]	m
	11	11.009 305 36(45)	0.801(7)		
6 C	12	12.0000000(00)	0.9893(8)	[12.0096, 12.0116]	
	13	13.003 354 835 07(23)	0.0107(8)		
	14	14.003 241 9884(40)			
7 N	14	14.003 074 004 43(20)	0.996 36(20)	[14.006 43, 14.007 28]	
	15	15.000 108 898 88(64)	0.003 64(20)		
8 O	16	15.994 914 619 57(17)	0.997 57(16)	[15.999 03, 15.999 77]	
	17	16.999 131 756 50(69)	0.000 38(1)		
	18	17.999 159 612 86(76)	0.002 05(14)		
9 F	19	18.998 403 162 73(92)	1	18.998 403 163(6)	
10 Ne	20	19.992 440 1762(17)	0.9048(3)	20.1797(6)	g,m
	21	20.993 846 685(41)	0.0027(1)		
	22	21.991 385 114(18)	0.0925(3)		

Figure C.1 Atomic Weights and Isotopic Compositions with Relative Atomic Masses for the first 10 elements (see National Institute of Standards and Technology (NIST) website NIST).

C.2 FUNDAMENTAL PHYSICAL CONSTANTS

NIST

2018 CODATA adjustment   From: **http://physics.nist.gov/constants**

Fundamental Physical Constants — Universal constants

Quantity	Symbol	Value	Unit	Relative std. uncert. u_r
speed of light in vacuum	c	299 792 458	m s^{-1}	exact
vacuum magnetic permeability $4\pi\alpha\hbar/e^2 c$	μ_0	1.256 637 062 12(19) × 10^{-6}	N A^{-2}	1.5 × 10^{-10}
$\mu_0/(4\pi \times 10^{-7})$		1.000 000 000 55(15)	N A^{-2}	1.5 × 10^{-10}
vacuum electric permittivity $1/\mu_0 c^2$	ϵ_0	8.854 187 8128(13) × 10^{-12}	F m^{-1}	1.5 × 10^{-10}
characteristic impedance of vacuum $\mu_0 c$	Z_0	376.730 313 668(57)	Ω	1.5 × 10^{-10}
Newtonian constant of gravitation	G	6.674 30(15) × 10^{-11}	m^3 kg^{-1} s^{-2}	2.2 × 10^{-5}
	$G/\hbar c$	6.708 83(15) × 10^{-39}	(GeV/c^2)$^{-2}$	2.2 × 10^{-5}
Planck constant*	h	6.626 070 15 × 10^{-34}	J Hz^{-1}	exact
		4.135 667 696... × 10^{-15}	eV Hz^{-1}	exact
	$\hbar$	1.054 571 817... × 10^{-34}	J s	exact
		6.582 119 569... × 10^{-16}	eV s	exact
	$\hbar c$	197.326 980 4...	MeV fm	exact
Planck mass $(\hbar c/G)^{1/2}$	m_P	2.176 434(24) × 10^{-8}	kg	1.1 × 10^{-5}
energy equivalent	$m_P c^2$	1.220 890(14) × 10^{19}	GeV	1.1 × 10^{-5}
Planck temperature $(\hbar c^5/G)^{1/2}/k$	T_P	1.416 784(16) × 10^{32}	K	1.1 × 10^{-5}
Planck length $\hbar/m_P c = (\hbar G/c^3)^{1/2}$	l_P	1.616 255(18) × 10^{-35}	m	1.1 × 10^{-5}
Planck time $l_P/c = (\hbar G/c^5)^{1/2}$	t_P	5.391 247(60) × 10^{-44}	s	1.1 × 10^{-5}

* The energy of a photon with frequency ν expressed in unit Hz is $E = h\nu$ in J. Unitary time evolution of the state of this photon is given by $\exp(-iEt/\hbar)|\varphi\rangle$, where $|\varphi\rangle$ is the photon state at time $t = 0$ and time is expressed in unit s. The ratio $Et/\hbar$ is a phase.

Figure C.2 Fundamental Physical Constants-Universal Constants (see National Institute of Standards and Technology (NIST) website, NIST).

C.3   PERIODIC TABLE

Periodic Table

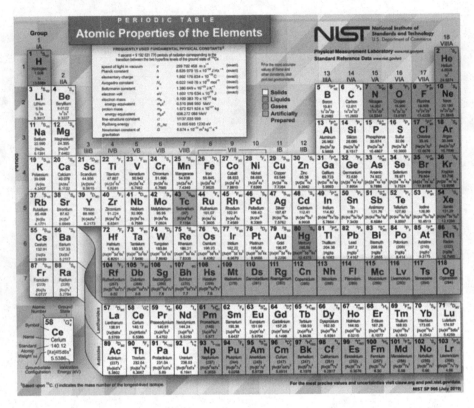

Figure C.3   Periodic Table: Atomic Properties of Elements (see National Institute of Standards and Technology (NIST) website, NIST).

C.4 PHOTON CROSS SECTIONS: XCOM

XCOM Developers and Contributors: M.J. Berger, J.H. Hubbell, S.M. Seltzer, J. Chang, J.S. Coursey, R. Sukumar, D.S. Zucker, and K. Olsen, NIST, PML, Radiation Physics Division. A web database is provided which can be used to calculate photon cross sections for scattering, photoelectric absorption and pair production, as well as total attenuation coefficients, for any element, compound or mixture (Z100), at energies from 1 keV to 100 GeV.

Constituents (Atomic Number : Fraction by Weight)

```
Z=1    : 0.111898
Z=8    : 0.888102
```

PHOTON ENERGY	SCATTERING COHERENT	INCOHER.	PHOTO- ELECTRIC ABSORPTION	PAIR PRODUCTION IN NUCLEAR FIELD	IN ELECTRON FIELD	TOTAL ATTENUATION WITH COHERENT SCATT.	WITHOUT COHERENT SCATT.
(MeV)	(cm2/g)	(cm2/g)	(cm2/g)	(cm2/g)	(cm2/g)	(cm2/g)	(cm2/g)
1.000E-03	1.372E+00	1.319E-02	4.076E+03	0.000E+00	0.000E+00	4.077E+03	4.076E+03
1.500E-03	1.269E+00	2.673E-02	1.374E+03	0.000E+00	0.000E+00	1.376E+03	1.374E+03
2.000E-03	1.150E+00	4.184E-02	6.162E+02	0.000E+00	0.000E+00	6.173E+02	6.162E+02
3.000E-03	9.087E-01	7.075E-02	1.919E+02	0.000E+00	0.000E+00	1.929E+02	1.919E+02
4.000E-03	7.082E-01	9.430E-02	8.197E+01	0.000E+00	0.000E+00	8.277E+01	8.207E+01
5.000E-03	5.579E-01	1.123E-01	4.192E+01	0.000E+00	0.000E+00	4.259E+01	4.203E+01
6.000E-03	4.489E-01	1.259E-01	2.407E+01	0.000E+00	0.000E+00	2.464E+01	2.419E+01
8.000E-03	3.102E-01	1.440E-01	9.919E+00	0.000E+00	0.000E+00	1.037E+01	1.006E+01
1.000E-02	2.205E-01	1.559E-01	4.944E+00	0.000E+00	0.000E+00	5.320E+00	5.099E+00
1.000E+00	5.627E-05	7.066E-02	3.681E-06	0.000E+00	0.000E+00	7.072E-02	7.066E-02
1.022E+00	5.388E-05	6.991E-02	3.430E-06	0.000E+00	0.000E+00	6.997E-02	6.991E-02
1.250E+00	3.603E-05	6.318E-02	2.329E-06	1.777E-05	0.000E+00	6.323E-02	6.320E-02
1.500E+00	2.501E-05	5.742E-02	1.690E-06	9.820E-05	0.000E+00	5.754E-02	5.752E-02
2.000E+00	1.407E-05	4.901E-02	1.063E-06	3.908E-04	0.000E+00	4.942E-02	4.940E-02
2.044E+00	1.347E-05	4.839E-02	1.028E-06	4.212E-04	0.000E+00	4.883E-02	4.881E-02
3.000E+00	6.255E-06	3.855E-02	5.937E-07	1.117E-03	1.349E-05	3.969E-02	3.968E-02
4.000E+00	3.519E-06	3.216E-02	4.075E-07	1.812E-03	5.507E-05	3.403E-02	3.402E-02
5.000E+00	2.252E-06	2.777E-02	3.090E-07	2.431E-03	1.097E-04	3.031E-02	3.031E-02
6.000E+00	1.564E-06	2.454E-02	2.484E-07	2.987E-03	1.685E-04	2.770E-02	2.770E-02
7.000E+00	1.149E-06	2.206E-02	2.075E-07	3.482E-03	2.272E-04	2.577E-02	2.577E-02
8.000E+00	8.796E-07	2.008E-02	1.780E-07	3.927E-03	2.843E-04	2.429E-02	2.429E-02
9.000E+00	6.951E-07	1.846E-02	1.559E-07	4.334E-03	3.389E-04	2.313E-02	2.313E-02
1.000E+01	5.630E-07	1.710E-02	1.386E-07	4.699E-03	3.910E-04	2.219E-02	2.219E-02
1.100E+01	4.652E-07	1.595E-02	1.248E-07	5.034E-03	4.404E-04	2.142E-02	2.142E-02
1.200E+01	3.909E-07	1.496E-02	1.134E-07	5.341E-03	4.872E-04	2.079E-02	2.079E-02
1.300E+01	3.331E-07	1.409E-02	1.039E-07	5.627E-03	5.315E-04	2.025E-02	2.025E-02
1.400E+01	2.872E-07	1.333E-02	9.591E-08	5.893E-03	5.737E-04	1.980E-02	1.980E-02
1.500E+01	2.502E-07	1.266E-02	8.906E-08	6.141E-03	6.135E-04	1.941E-02	1.941E-02
1.600E+01	2.199E-07	1.205E-02	8.311E-08	6.376E-03	6.519E-04	1.908E-02	1.908E-02
1.800E+01	1.737E-07	1.102E-02	7.331E-08	6.803E-03	7.230E-04	1.854E-02	1.854E-02
2.000E+01	1.407E-07	1.016E-02	6.555E-08	7.186E-03	7.878E-04	1.813E-02	1.813E-02

Figure C.4 Photon Cross Sections for Water (as an example) calculated using the XCOM Program (see National Institute of Standards and Technology (NIST) website, XCOM).

C.5 STOPPING-POWER & RANGE TABLES FOR ELECTRONS: ESTAR

ESTAR Developers and Contributors: M.J. Berger, J.S. Coursey, M.A. Zucker and J. Chang, National Institute of Standards and Technology (NIST), Physics Laboratory (now Physical Measurement Laboratory (PML) and Office of Electronic Commerce in Scientific and Engineering Data. The databases ESTAR, PSTAR and ASTAR calculate stopping-power and range tables for electrons, protons, or helium ions, according to methods described in ICRU Reports 37 and 49. Stopping-power and range tables can be calculated for electrons in any user-specified material and for protons and helium ions in 74 materials.

(required) Kinetic Energy (MeV)	Stopping Power (MeV cm²/g)			CSDA Range (g/cm²)	Radiation Yield	Density Effect Parameter
	Collision	Radiative	Total			
1.000E-02	2.256E+01	3.898E-03	2.256E+01	2.515E-04	9.408E-05	0.000E+00
1.250E-02	1.897E+01	3.927E-03	1.898E+01	3.728E-04	1.133E-04	0.000E+00
1.500E-02	1.647E+01	3.944E-03	1.647E+01	5.147E-04	1.316E-04	0.000E+00
1.750E-02	1.461E+01	3.955E-03	1.461E+01	6.762E-04	1.493E-04	0.000E+00
2.000E-02	1.317E+01	3.963E-03	1.318E+01	8.566E-04	1.663E-04	0.000E+00
2.500E-02	1.109E+01	3.974E-03	1.110E+01	1.272E-03	1.990E-04	0.000E+00
3.000E-02	9.653E+00	3.984E-03	9.657E+00	1.756E-03	2.301E-04	0.000E+00
3.500E-02	8.592E+00	3.994E-03	8.596E+00	2.306E-03	2.599E-04	0.000E+00
4.000E-02	7.777E+00	4.005E-03	7.781E+00	2.919E-03	2.886E-04	0.000E+00
4.500E-02	7.130E+00	4.018E-03	7.134E+00	3.591E-03	3.165E-04	0.000E+00
5.000E-02	6.603E+00	4.031E-03	6.607E+00	4.320E-03	3.435E-04	0.000E+00
5.500E-02	6.166E+00	4.046E-03	6.170E+00	5.103E-03	3.698E-04	0.000E+00
6.000E-02	5.797E+00	4.062E-03	5.801E+00	5.940E-03	3.955E-04	0.000E+00
7.000E-02	5.207E+00	4.098E-03	5.211E+00	7.762E-03	4.453E-04	0.000E+00
8.000E-02	4.757E+00	4.138E-03	4.761E+00	9.773E-03	4.931E-04	0.000E+00
9.000E-02	4.402E+00	4.181E-03	4.407E+00	1.196E-02	5.393E-04	0.000E+00
1.000E-01	4.115E+00	4.228E-03	4.119E+00	1.431E-02	5.842E-04	0.000E+00
1.250E-01	3.591E+00	4.355E-03	3.596E+00	2.083E-02	6.912E-04	0.000E+00
1.500E-01	3.238E+00	4.494E-03	3.242E+00	2.817E-02	7.926E-04	0.000E+00
1.750E-01	2.984E+00	4.643E-03	2.988E+00	3.622E-02	8.894E-04	0.000E+00
2.000E-01	2.793E+00	4.801E-03	2.798E+00	4.488E-02	9.826E-04	0.000E+00
2.500E-01	2.528E+00	5.141E-03	2.533E+00	6.372E-02	1.161E-03	0.000E+00
3.000E-01	2.355E+00	5.514E-03	2.360E+00	8.421E-02	1.331E-03	0.000E+00
3.500E-01	2.235E+00	5.914E-03	2.241E+00	1.060E-01	1.496E-03	0.000E+00
4.000E-01	2.148E+00	6.339E-03	2.154E+00	1.288E-01	1.658E-03	0.000E+00
4.500E-01	2.083E+00	6.787E-03	2.090E+00	1.523E-01	1.818E-03	0.000E+00
5.000E-01	2.034E+00	7.257E-03	2.041E+00	1.765E-01	1.976E-03	0.000E+00
5.500E-01	1.995E+00	7.747E-03	2.003E+00	2.013E-01	2.134E-03	1.103E-02
6.000E-01	1.963E+00	8.254E-03	1.972E+00	2.265E-01	2.292E-03	2.938E-02
7.000E-01	1.917E+00	9.313E-03	1.926E+00	2.778E-01	2.608E-03	7.435E-02

Figure C.5 Electronic Stopping Power and Range Table for liquid water (as an example) calculated using the ESTAR (see National Institute of Standards and Technology (NIST) website, ESTAR).

C.6 STOPPING-POWER & RANGE TABLES FOR PROTONS: PSTAR

PSTAR Developers and Contributors: M.J. Berger, J.S. Coursey, M.A. Zucker and J. Chang, National Institute of Standards and Technology (NIST), Physics Laboratory (now Physical Measurement Laboratory (PML) and Office of Electronic Commerce in Scientific and Engineering Data. The databases ESTAR, PSTAR and ASTAR calculate stopping-power and range tables for electrons, protons, or helium ions, according to methods described in ICRU Reports 37 and 49. Stopping-power and range tables can be calculated for electrons in any user-specified material and for protons and helium ions in 74 materials.

(required) Kinetic Energy (MeV)	Stopping Power (MeV cm²/g)			Range		
	Electronic	Nuclear	Total	CSDA (g/cm²)	Projected (g/cm²)	Detour Factor Projected / CSDA
1.000E-03	1.337E+02	4.315E+01	1.769E+02	6.319E-06	2.878E-06	0.4555
1.500E-03	1.638E+02	3.460E+01	1.984E+02	8.969E-06	4.400E-06	0.4906
2.000E-03	1.891E+02	2.927E+01	2.184E+02	1.137E-05	5.909E-06	0.5197
2.500E-03	2.114E+02	2.557E+01	2.370E+02	1.357E-05	7.380E-06	0.5440
3.000E-03	2.316E+02	2.281E+01	2.544E+02	1.560E-05	8.811E-06	0.5647
4.000E-03	2.675E+02	1.894E+01	2.864E+02	1.930E-05	1.155E-05	0.5986
5.000E-03	2.990E+02	1.631E+01	3.153E+02	2.262E-05	1.415E-05	0.6254
6.000E-03	3.276E+02	1.439E+01	3.420E+02	2.567E-05	1.661E-05	0.6473
7.000E-03	3.538E+02	1.292E+01	3.667E+02	2.849E-05	1.896E-05	0.6656
8.000E-03	3.782E+02	1.175E+01	3.900E+02	3.113E-05	2.121E-05	0.6813
9.000E-03	4.012E+02	1.080E+01	4.120E+02	3.363E-05	2.337E-05	0.6950
1.000E-02	4.229E+02	1.000E+01	4.329E+02	3.599E-05	2.545E-05	0.7070
1.250E-02	4.660E+02	8.485E+00	4.745E+02	4.150E-05	3.037E-05	0.7318
1.500E-02	5.036E+02	7.400E+00	5.110E+02	4.657E-05	3.499E-05	0.7514
1.750E-02	5.372E+02	6.581E+00	5.437E+02	5.131E-05	3.938E-05	0.7674
2.000E-02	5.675E+02	5.939E+00	5.733E+02	5.578E-05	4.356E-05	0.7808
2.250E-02	5.946E+02	5.421E+00	6.001E+02	6.005E-05	4.757E-05	0.7923
2.500E-02	6.195E+02	4.993E+00	6.245E+02	6.413E-05	5.144E-05	0.8022
2.750E-02	6.421E+02	4.633E+00	6.467E+02	6.806E-05	5.519E-05	0.8109
3.000E-02	6.628E+02	4.325E+00	6.671E+02	7.187E-05	5.883E-05	0.8187
3.500E-02	6.989E+02	3.826E+00	7.028E+02	7.916E-05	6.585E-05	0.8319
4.000E-02	7.290E+02	3.437E+00	7.324E+02	8.613E-05	7.259E-05	0.8429
4.500E-02	7.538E+02	3.126E+00	7.569E+02	9.284E-05	7.911E-05	0.8522
5.000E-02	7.740E+02	2.870E+00	7.768E+02	9.935E-05	8.547E-05	0.8602
5.500E-02	7.901E+02	2.655E+00	7.927E+02	1.057E-04	9.169E-05	0.8673
6.000E-02	8.026E+02	2.473E+00	8.050E+02	1.120E-04	9.782E-05	0.8735
6.500E-02	8.119E+02	2.316E+00	8.147E+02	1.182E-04	1.039E-04	0.8791
7.000E-02	8.183E+02	2.178E+00	8.205E+02	1.243E-04	1.099E-04	0.8842
7.500E-02	8.223E+02	2.058E+00	8.243E+02	1.303E-04	1.159E-04	0.8889

Figure C.6 Stopping Power and Range Table for Protons in liquid water (as an example) calculated using the PSTAR (see National Institute of Standards and Technology (NIST) website, PSTAR).

Bibliography

[1] Steven P. Ahlen. Theoretical and experimental aspects of the energy loss of relativistic heavily ionizing particles. *Reviews of Modern Physics*, 52(1):121–173, Jan 1980.

[2] A. Andronic and J. P. Wessels. Transition Radiation Detectors. *arXiv:1111.4188 [physics.ins-det]*, Nov 2011.

[3] Frank Herbert Attix. *Introduction to Radiological Physics and Radiation Dosimetry*. Wiley, Nov 1986.

[4] Walter H. Barkas, Wallace Birnbaum, and Frances M. Smith. Mass-Ratio Method Applied to the Measurement of L -Meson Masses and the Energy Balance in Pion Decay. *Physical Review*, 101(2):778–795, Jan 1956.

[5] Roger (Roger J.) Barlow. *Statistics: a guide to the use of statistical methods in the physical sciences*. Wiley, 1989.

[6] Harry Bateman. Solution of a system of differential equations occurring in the theory of radioactive transformations. *Proceedings of the Cambridge Philosophical Society*, 15:423–427, 1910.

[7] M. J. Berger, M. Inokuti, H. H. Andersen, H. Bichsel, D. Powers, S. M. Seltzer, D. Thwaites, and D. E. Watt. ICRU Report 49. *Journal of the International Commission on Radiation Units and Measurements*, os25(2), 1993.

[8] M. J. Berger, M. Inokuti, H. H. Anderson, H. Bichsel, J. A. Dennis, D. Powers, S. M. Seltzer, and J. E. Turner. ICRU Report 37. Stopping Powers for Electrons and Positrons. *Journal of the International Commission on Radiation Units and Measurements*, os19(2), Dec 1984.

[9] Martin J. Berger and Ruqing Wang. Multiple-Scattering Angular Deflections and Energy- Loss Straggling. In *Monte Carlo Transport of Electrons and Photons*, pages 21–56. Springer US, 1988.

[10] J. Beringer, J. F. Arguin, R. M. Barnett, K. Copic, O. Dahl, D. E. Groom, C. J. Lin, J. Lys, H. Murayama, C. G. Wohl, W. M. Yao, P. A. Zyla, C. Amsler, M. Antonelli, D. M. Asner, H. Baer, H. R. Band, T. Basaglia, C. W. Bauer, J. J. Beatty, V. I. Belousov, E. Bergren, G. Bernardi, W. Bertl, S. Bethke, H. Bichsel, O. Biebel, E. Blucher, S. Blusk, G. Brooijmans, O. Buchmueller, R. N. Cahn, M. Carena, A. Ceccucci, D. Chakraborty, M. C. Chen, R. S. Chivukula, G. Cowan, G. D'ambrosio, T. Damour, D. De Florian, A. De Gouvêa, T. Degrand, P. De Jong, G. Dissertori, B. Dobrescu, M. Doser, M. Drees, D. A. Edwards, S. Eidelman, J. Erler, V. V. Ezhela, W. Fetscher, B. D. Fields, B. Foster, T. K. Gaisser, L. Garren, H. J. Gerber, G. Gerbier, T. Gherghetta, S. Golwala, M. Goodman, C. Grab, A. V. Gritsan, J. F. Grivaz, M. Grünewald, A. Gurtu, T. Gutsche, H. E. Haber, K. Hagiwara, C. Hagmann, C. Hanhart, S. Hashimoto, K. G. Hayes, M. Heffner, B. Heltsley, J. J. Hernández-Rey, K. Hikasa, A. Höcker, J. Holder, A. Holtkamp, J. Huston, J. D. Jackson, K. F. Johnson, T. Junk, D. Karlen, D. Kirkby, S. R. Klein, E. Klempt, R. V. Kowalewski, F. Krauss, M. Kreps, B. Krusche, Y. V. Kuyanov, Y. Kwon, O. Lahav, J. Laiho, P. Langacker, A. Liddle, Z. Ligeti, T. M. Liss, L. Littenberg, K. S. Lugovsky, S. B. Lugovsky, T. Mannel, A. V. Manohar, W. J. Marciano, A. D. Martin, A. Masoni, J. Matthews, D. Milstead, R. Miquel, K. Mönig, F. Moortgat, K. Nakamura, M. Narain, P. Nason, S. Navas, M. Neubert, P. Nevski, Y. Nir, K. A. Olive, L. Pape, J. Parsons, C. Patrignani, J. A. Peacock, S. T. Petcov, A. Piepke, A. Pomarol, G. Punzi, A. Quadt, S. Raby, G. Raffelt, B. N. Ratcliff, P. Richardson, S. Roesler, S. Rolli, A. Romaniouk, L. J. Rosenberg, J. L. Rosner, C. T. Sachrajda, Y. Sakai, G. P. Salam, S. Sarkar, F. Sauli, O. Schneider, K. Scholberg, D. Scott, W. G. Seligman, M. H. Shaevitz, S. R. Sharpe, M. Silari, T. Sjöstrand, P. Skands, J. G. Smith, G. F. Smoot, S. Spanier, H. Spieler, A. Stahl, T. Stanev, S. L. Stone, T. Sumiyoshi, M. J. Syphers, F. Takahashi, M. Tanabashi, J. Terning, M. Titov, N. P. Tkachenko, N. A. Törnqvist, D. Tovey, G. Valencia, K. Van Bibber, G. Venanzoni, M. G. Vincter, P. Vogel, A. Vogt, W. Walkowiak, C. W. Walter, D. R. Ward, T. Watari, G. Weiglein, E. J. Weinberg,

L. R. Wiencke, L. Wolfenstein, J. Womersley, C. L. Woody, R. L. Workman, A. Yamamoto, G. P. Zeller, O. V. Zenin, J. Zhang, R. Y. Zhu, G. Harper, V. S. Lugovsky, and P. Schaffner. Review of particle physics. *Physical Review D - Particles, Fields, Gravitation and Cosmology*, 86(1), Jul 2012.

[11] H. Bethe. Zur Theorie des Durchgangs schneller Korpuskularstrahlen durch Materie. *Annalen der Physik*, 397(3):325–400, Jan 1930.

[12] H. Bethe. Bremsformel für Elektronen relativistischer Geschwindigkeit. *Zeitschrift für Physik*, 76(5-6):293–299, May 1932.

[13] H. Bethe. Theory of the Passage of Fast Corpuscular Rays Through Matter. Pages 77–154. Jul 1997.

[14] H. Bethe and W. Heitler. On the Stopping of Fast Particles and on the Creation of Positive Electrons. *Proceedings of the Royal Society A: Mathematical, Physical and Engineering Sciences*, 146(856):83–112, Aug 1934.

[15] H. A. Bethe. Molière's theory of multiple scattering. *Physical Review*, 89(6):1256–1266, 1953.

[16] H. Bichsel. Barkas effect and effective charge in the theory of stopping power. *Physical Review. A, Atomic, molecular, and optical physics*, 41(7):3642–3647, Apr 1990.

[17] F. Bloch. Zur Bremsung rasch bewegter Teilchen beim Durchgang durch Materie. *Annalen der Physik*, 408(3):285–320, Jan 1933.

[18] N. Bohr. II. <i>On the theory of the decrease of velocity of moving electrified particles on passing through matter</i>. *The London, Edinburgh, and Dublin Philosophical Magazine and Journal of Science*, 25(145):10–31, Jan 1913.

[19] Niels Bohr. On the Constitution of Atoms and Molecules, {Part I}. {The} binding of electrons by positive nuclei. *Philosophical Magazine*, 26:1–25, 1913.

[20] Max Born. Zur Quantenmechanik der Stossvorgange. *Zeitschrift fur Physik*, 37(12):863–867, Dec 1926.

[21] W. H. Bragg and R. Kleeman. XXXIX. On the α particles of radium, and their loss of range in passing through various atoms and molecules. *The London, Edinburgh, and Dublin Philosophical Magazine and Journal of Science*, 10(57):318–340, Sep 1905.

[22] Marco Capogni, Antonino Pietropaolo, Lina Quintieri, Maurizio Angelone, Alessandra Boschi, Mauro Capone, Nadia Cherubini, Pierino De Felice, Alessandro Dodaro, Adriano Duatti, Aldo Fazio, Stefano Loreti, Petra Martini, Guglielmo Pagano, Micol Pasquali, Mario Pillon, Licia Uccelli, and Aldo Pizzuto. 14 MeV Neutrons for 99Mo/99mTc Production: Experiments, Simulations and Perspectives. *Molecules*, 23(8):1872, Jul 2018.

[23] Herman. Cember and Thomas E. (Thomas Edward) Johnson. *Introduction to health physics*. McGraw-Hill Medical, 2009.

[24] Jerzy Cetnar. General solution of Bateman equations for nuclear transmutations. *Annals of Nuclear Energy*, 33(7):640–645, 2006.

[25] Sow-Hsin Chen and Michael. Kotlarchyk. *Interactions of photons and neutrons with matter*. World Scientific, 2007.

[26] Simon R. Cherry, James A. Sorenson, and Michael E. Phelps. *Physics in nuclear medicine*. Elsevier/Saunders, 2012.

[27] Esther Ciarrocchi and Nicola Belcari. Cerenkov luminescence imaging: physics principles and potential applications in biomedical sciences. *EJNMMI Physics*, 4(1), Dec 2017.

[28] Arthur H. Compton. A Quantum Theory of the Scattering of X-rays by Light Elements. *Physical Review*, 21(5):483–502, May 1923.

[29] P. A. M. Dirac. The Quantum Theory of the Emission and Absorption of Radiation. *Proceedings of the Royal Society A: Mathematical, Physical and Engineering Sciences*, 114(767):243–265, Mar 1927.

[30] P. A. M. Dirac. The Quantum Theory of the Electron. *Proceedings of the Royal Society A: Mathematical, Physical and Engineering Sciences*, 117(778):610–624, Feb 1928.

[31] P. A. M. Dirac. The Quantum Theory of the Electron. Part II. *Proceedings of the Royal Society A: Mathematical, Physical and Engineering Sciences*, 118(779):351–361, Mar 1928.

[32] U Fano. Penetration of Protons, Alpha Particles, and Mesons. *Annual Review of Nuclear Science*, 13(1):1–66, Dec 1963.

[33] Ian Farrance and Robert Frenkel. Uncertainty in measurement: a review of Monte Carlo simulation using microsoft excel for the calculation of uncertainties through functional relationships, including uncertainties in empirically derived constants. *The Clinical Biochemist Reviews*, 35(1):37–61, Feb 2014.

[34] Gary J. Feldman and Robert D. Cousins. A Unified Approach to the Classical Statistical Analysis of Small Signals. *arXiv:physics/9711021 [physics.data-an]*, Nov 1997.

[35] E. Fermi. Versuch einer Theorie der β-Strahlen. I. *Zeitschrift für Physik*, 88(3-4):161–177, Mar 1934.

[36] Enrico Fermi. Quantum theory of radiation. *Reviews of Modern Physics*, 4(1):87–132, Jan 1932.

[37] Enrico Fermi. Tentativo di una Teoria Dei Raggi β. *Il Nuovo Cimento*, 11(1):1–19, Jan 1934.

[38] Enrico Fermi. The Ionization Loss of Energy in Gases and in Condensed Materials. *Physical Review*, 57(6):485–493, Mar 1940.

[39] G. Gamow. Zur Quantentheorie des Atomkernes. *Zeitschrift fur Physik*, 51(3-4):204–212, Mar 1928.

[40] H. Geiger and J.M. Nuttall. LVII. The ranges of the α particles from various radioactive substances and a relation between range and period of transformation. *The London, Edinburgh, and Dublin Philosophical Magazine and Journal of Science*, 22(130):613–621, Oct 1911.

[41] H. Geiger and J.M. Nuttall. XL. The ranges of the α particles from uranium. *The London, Edinburgh, and Dublin Philosophical Magazine and Journal of Science*, 23(135):439–445, Mar 1912.

[42] Spyridon Gkoumas, Zhentian Wang, Matteo Abis, Carolina Arboleda, George Tudosie, Tilman Donath, Christian Brönnimann,

Clemens Schulze-Briese, and Marco Stampanoni. Grating-based interferometry and hybrid photon counting detectors: Towards a new era in X-ray medical imaging. *Nuclear Instruments and Methods in Physics Research, Section A: Accelerators, Spectrometers, Detectors and Associated Equipment*, 809:23–30, Feb 2016.

[43] David J. Griffiths and Darrell F. Schroeter. *Introduction to Quantum Mechanics*. Cambridge University Press, 3rd edition, Aug 2018.

[44] Eric J. Hall and Amato J. Giaccia. *Radiobiology for the Radiologist*, volume 8. Wolters Kluwer – Medknow Publications, Philadelphia, 8th edition, Nov 2018.

[45] H E Hansen and U Ingerslev-Jensen. Penetration of fast electrons and positrons. *Journal of Physics D: Applied Physics*, 16(7):1353–1370, Jul 1983.

[46] Hiroaki Hayashi, Yasuaki Kojima, Michihiro Shibata, and Kiyoshi Kawade. Practical $Q\beta$ analysis method based on the Fermi-Kurie plot for spectra measured with total absorption BGO detector. *Nuclear Instruments and Methods in Physics Research, Section A: Accelerators, Spectrometers, Detectors and Associated Equipment*, 613(1):79–89, Jan 2010.

[47] W. Heitler. *Quantum Theory of Radiation*. Oxford University Press, Oxford, 1954.

[48] W. Heitler and F. Sauter. Stopping of Fast Particles with Emission of Radiation and the Birth of Positive Electrons. *Nature*, 132(3345):892–892, Dec 1933.

[49] K. Heyde. *Basic Ideas and Concepts in Nuclear Physics : An Introductory Approach*. Institute of Physics, Taylor and Francis, 3rd edition, 2004.

[50] J. H. Hubbell. Review of photon interaction cross section data in the medical and biological context. *Physics in Medicine and Biology*, 44(1):1–22, 1999.

[51] Miljenko Huzak. Chi-Square Distribution. In *International Encyclopedia of Statistical Science*, pages 245–246. Springer Berlin Heidelberg, 2011.

[52] IAEA. *Nuclear Medicine Physics: A Handbook for Teachers and Students*. IAEA, Vienna, 2014.

[53] John David Jackson. *Classical Electrodynamics*. Wiley, 3rd edition, 2001.

[54] Frederick James. *Statistical Methods in Experimental Physics*. World Scientific, Nov 2006.

[55] Alan Jeffrey and Hui-Hui Dai. *Handbook of Mathematical Formulas and Integrals*. Elsevier, Amsterdam, London, New York, 4th edition, 2008.

[56] David Jette and Suzan Walker. Electron dose calculation using multiple-scattering theory: Energy distribution due to multiple scattering. *Medical Physics*, 24(3):383–400, 1997.

[57] Toni Kaltiaisenaho. Photon transport physics in Serpent 2 Monte Carlo code. *Computer Physics Communications*, 252, Jul 2020.

[58] L. Katz and A. S. Penfold. Range-energy relations for electrons and the determination of beta-ray end-point energies by absorption. *Reviews of Modern Physics*, 24(1):28–44, 1952.

[59] O. Klein and Y. Nishina. The Scattering of Light by Free Electrons according to Dirac's New Relativistic Dynamics. *Nature*, 122(3072):398–399, Sep 1928.

[60] O. Klein and Y. Nishina. Über die Streuung von Strahlung durch freie Elektronen nach der neuen relativistischen Quantendynamik von Dirac. *Zeitschrift für Physik*, 52(11-12):853–868, Nov 1929.

[61] Glenn F Knoll. *Radiation Detection and Measurement*. John Wiley & Sons, Inc., 4th edition, 2010.

[62] Michael Kotlarchyk and Sow-hsin Chen. *Interactions Of Photons And Neutrons With Matter (2nd Edition)*. World Scientific Publishing Company, 2000.

[63] Richard E. Kozack. Born approximation and differential cross sections in nuclear physics. *American Journal of Physics*, 59(1):74–79, Jan 1991.

[64] Kenneth Krane. *Introductory nuclear physics*. Wiley, New York, 1988.

[65] F. N. D. Kurie. On the Use of the Kurie Plot. *Physical Review*, 73(10):1207–1207, May 1948.

[66] William R. Leo. Statistics and the Treatment of Experimental Data. *Techniques for Nuclear and Particle Physics Experiments*, pages 75–106, 1987.

[67] W. P. Levin, H. Kooy, J. S. Loeffler, and T. F. DeLaney. Proton beam therapy, Oct 2005.

[68] Edmond Levy. A matrix exponential approach to radioactive decay equations. *American Journal of Physics*, 86(12):909–913, Dec 2018.

[69] Jens Lindhard and Allan H. Sorensen. Relativistic theory of stopping for heavy ions. *Physical Review A*, 53(4):2443–2456, Apr 1996.

[70] R. A. Lloyd. Application of the Kurie plot to the standardization of pure beta-emitters. *Nature*, 181(4615):1055–1056, 1958.

[71] Renata Longo. Current studies and future perspectives of synchrotron radiation imaging trials in human patients. *Nuclear Instruments and Methods in Physics Research, Section A: Accelerators, Spectrometers, Detectors and Associated Equipment*, 809:13–22, Feb 2016.

[72] Renata Longo, Fulvia Arfelli, Deborah Bonazza, Ubaldo Bottigli, Luca Brombal, Adriano Contillo, Maria A. Cova, Pasquale Delogu, Francesca Di Lillo, Vittorio Di Trapani, Sandro Donato, Diego Dreossi, Viviana Fanti, Christian Fedon, Bruno Golosio, Giovanni Mettivier, Piernicola Oliva, Serena Pacilè, Antonio Sarno, Luigi Rigon, Paolo Russo, Angelo Taibi, Maura Tonutti, Fabrizio Zanconati, and Giuliana Tromba. Advancements towards the implementation of clinical phase-contrast breast computed tomography at Elettra. *Journal of Synchrotron Radiation*, 26:1343–1353, Jul 2019.

[73] Louis Lyons. *Statistics for Nuclear and Particle Physicists*. Cambridge University Press, Mar 1986.

[74] C Ma and A E Nahum. Bragg-Gray theory and ion chamber dosimetry for photon beams. *Physics in Medicine & Biology*, 1991.

[75] Xiaowei Ma, Jing Wang, and Zhen Cheng. Cerenkov radiation: a multi-functional approach for biological sciences. *Frontiers in Physics*, 2, 2014.

[76] Carlos Mana. Probability and Statistics for Particle Physics. *Springer*, 2017.

[77] Silvia Masciocchi and Gsi Darmstadt. Statistical Methods in Particle Physics, 2013.

[78] Gert Moliere. Theorie der Streuung schneller geladener Teilchen II Mehrfach-und Vielfachstreuung1. *Zeitschrift fur Naturforschung - Section A Journal of Physical Sciences*, 3(2):78–97, Feb 1948.

[79] Atsushi Momose, Wataru Yashiro, Kazuhiro Kido, Junko Kiyohara, Chiho Makifuchi, Tsukasa Ito, Sumiya Nagatsuka, Chika Honda, Daiji Noda, Tadashi Hattori, Tokiko Endo, Masabumi Nagashima, and Junji Tanaka. X-ray phase imaging: From synchrotron to hospital. *Philosophical Transactions of the Royal Society A: Mathematical, Physical and Engineering Sciences*, 372(2010), Mar 2014.

[80] Peter Monk. *Finite Element Methods for Maxwell's Equations.* Oxford University Press, Apr 2003.

[81] L. Moral and A. F. Pacheco. Algebraic approach to the radioactive decay equations. *American Journal of Physics*, 71(7):684–686, Jul 2003.

[82] Paweł Moskal, Bożena Jasińska, Ewa Stepień, and Steven D. Bass. Positronium in medicine and biology. *Nature Reviews Physics*, 1(9):527–529, Sep 2019.

[83] Toshio Namba. Precise measurement of positronium. *Progress of Theoretical and Experimental Physics*, 2012(1):04D003, 2012.

[84] Jacob Neufeld. Bohr's Theory of Energy Losses of Moving Charged Particles. *Physical Review*, 95(5):1128–1133, Sep 1954.

[85] H. Nikjoo, S. Uehara, D. Emfietzoglou, and F. A. Cucinotta. Track-structure codes in radiation research. *Radiation Measurements*, 41(9-10):1052–1074, Oct 2006.

[86] L. C. Northcliffe and R. F. Schilling. Range and stopping-power tables for heavy ions. *Atomic Data and Nuclear Data Tables*, 7(3-4):233–463, Jan 1970.

[87] Hywel Owen, Antony Lomax, and Simon Jolly. Current and future accelerator technologies for charged particle therapy. *Nuclear Instruments and Methods in Physics Research, Section A: Accelerators, Spectrometers, Detectors and Associated Equipment*, 809:96–104, 2016.

[88] Daniele Panetta. Advances in X-ray detectors for clinical and preclinical Computed Tomography. *Nuclear Instruments and Methods in Physics Research, Section A: Accelerators, Spectrometers, Detectors and Associated Equipment*, 809:2–12, Feb 2016.

[89] Karl Pearson. X. On the criterion that a given system of deviations from the probable in the case of a correlated system of variables is such that it can be reasonably supposed to have arisen from random sampling. *The London, Edinburgh, and Dublin Philosophical Magazine and Journal of Science*, 50(302):157–175, Jul 1900.

[90] Yoav Peleg, Reuven Pnini, and Elyahu Zaarur. Schaum's Outline of Theory and Problems of Quantum Mechanics, 1998.

[91] B Y Randolph S Peterson. *Experimental γ Ray Spectroscopy and Investigations of Environmental Radioactivity*. Spectrum Techniques, 1998.

[92] Ervin B. Podgoršak. *Compendium to Radiation Physics for Medical Physicists: 300 Problems and Solutions*. Springer Berlin Heidelberg, 2014.

[93] S Pommé. The uncertainty of the half-life. *Metrologia*, 52(3):S51–S65, Jun 2015.

[94] R. H. Pratt, Akiva Ron, and H. K. Tseng. Atomic photoelectric effect above 10 keV, Apr 1973.

[95] Dobromir S. Pressyanov. Short solution of the radioactive decay chain equations. *American Journal of Physics*, 70(4):444–445, 2002.

[96] C. Qi, A. N. Andreyev, M. Huyse, R. J. Liotta, P. Van Duppen, and R. Wyss. On the validity of the Geiger-Nuttall alpha-decay law and its microscopic basis. *Physics Letters, Section B: Nuclear, Elementary Particle and High-Energy Physics*, 734:203–206, Jun 2014.

[97] Chong Qi, R J Liotta, and R Wyss. Generalization of the Geiger-Nuttall law and alpha clustering in heavy nuclei. *Journal of Physics: Conference Series*, 381(1):012131, Sep 2012.

[98] Chong Qi, Roberto Liotta, and Ramon Wyss. Recent developments in radioactive charged-particle emissions and related phenomena. *Progress in Particle and Nuclear Physics*, 105:214–251, Mar 2019.

[99] Ashwani Rajan and Shantanu Desai. A meta-analysis of neutron lifetime measurements. Dec 2018.

[100] D. W. O. Rogers and Reid W. Townson. On calculating kerma, collision kerma and radiative yields. *Medical Physics*, 46(11):5173–5184, Nov 2019.

[101] J. A. Rowlands. The physics of computed radiography. *Physics in Medicine and Biology*, 47(23), Dec 2002.

[102] E. Rutherford. LXXIX. The scattering of α and β particles by matter and the structure of the atom. *The London, Edinburgh, and Dublin Philosophical Magazine and Journal of Science*, 21(125):669–688, May 1911.

[103] Lorenzo Sabbatucci and Francesc Salvat. *Radiation Physics and Chemistry*, 121:122–140, Apr 2016.

[104] Fritz Sauter. Über den atomaren Photoeffekt in der K-Schale nach der relativistischen Wellenmechanik Diracs. *Annalen der Physik*, 403(4):454–488, Jan 1931.

[105] R. A. Schapery and S.W. Park. Methods of interconversion between linear viscoelastic material functions. Part II—an approximate analytical method. *Int J Solids Struct*, 36(11):1677–1699, Apr 1999.

[106] E. Schrödinger. An Undulatory Theory of the Mechanics of Atoms and Molecules. *Phys. Rev.*, 28(6):1049–1070, Dec 1926.

[107] J. H. Scofield. Theoretical photoionization cross sections from 1 to 1500 keV. Technical report, 1973.

[108] SM Seltzer, J M Fernandez-Varea, P Andreo, P M Bergstrom, D T Burns, I Krajcar Bronic, CK Ross, and F Salvat. *ICRU Report 90. Key Data for Ionizing-Radiation Dosimetry: Measurement Standards and Application*, volume 14. ICRU, 2014.

[109] Stephen M. Seltzer and Martin J. Berger. Evaluation of the collision stopping power of elements and compounds for electrons and positrons. *The International Journal of Applied Radiation and Isotopes*, 33(11):1189–1218, Nov 1982.

[110] Ramamurti Shankar. *Principles of quantum mechanics*. Plenum Press, New York, 2nd edition, 1994.

[111] Kengo Shibuya, Haruo Saito, Fumihiko Nishikido, Miwako Takahashi, and Taiga Yamaya. Oxygen sensing ability of positronium atom for tumor hypoxia imaging. *Communications Physics*, 3(1), Dec 2020.

[112] J.W. Strutt. XV. On the light from the sky, its polarization and colour. *The London, Edinburgh, and Dublin Philosophical Magazine and Journal of Science*, 41(271):107–120, Feb 1871.

[113] Michael P H Stumpf and Mason A Porter. Critical truths about power laws. *Science*, 335(6069):665–6, Feb 2012.

[114] D. Tan and B. Heaton. Simple empirical relations for electron CSDA range and electron energy loss. *Applied Radiation and Isotopes*, 45(4):527–528, 1994.

[115] John R Taylor. *An Introduction to Error Analysis*. University Science Books, Sausalito, CA, 2nd edition, 1997.

[116] David J. Thomas. ICRU report 85: fundamental quantities and units for ionizing radiation. *Radiation Protection Dosimetry*, 150(4):550–552, Jul 2012.

[117] R. G. Thomas. A Formulation of the Theory of Alpha-particle Decay from Time-independent Equations. *Progress of Theoretical Physics*, 12(3):253–264, Sep 1954.

[118] D. I. Thwaites. Bragg's Rule of Stopping Power Additivity: A Compilation and Summary of Results. *Radiation Research*, 95(3):495, Sep 1983.

[119] Yung Su Tsai. Pair production and bremsstrahlung of charged leptons. *Reviews of Modern Physics*, 46(4):815–851, 1974.

[120] Yung Su Tsai. Erratum: Pair production and bremsstrahlung of charged leptons (Reviews of Modern Physics), 1977.

[121] Attila Vértes. *Handbook of nuclear chemistry*. Springer, 2011.

[122] M. A. Waggoner. Radioactive decay of Cs137. *Physical Review*, 82(6):906–909, 1951.

[123] B A Weaver and A J Westphal. Energy loss of relativistic heavy ions in matter. *Nuclear Instruments and Methods in Physics Research Section B: Beam Interactions with Materials and Atoms*, 187(3):285–301, 2002.

[124] Eric W. Weisstein. Chi-Squared Distribution.

[125] Patricia Wells. *Practical Mathematics in Nuclear Medicine Technology*. 2 edition, 2011.

[126] Fred L. Wilson. Fermi's Theory of Beta Decay. *American Journal of Physics*, 36(12):1150–1160, Dec 1968.

[127] Yuji Yazaki. How the Klein-Nishina formula was derived: Based on the Sangokan Nishina Source Materials. *Proceedings of the Japan Academy. Series B, Physical and biological sciences*, 93(6):399 421, 2017.

[128] J. F. Ziegler. Stopping of energetic light ions in elemental matter. *Journal of Applied Physics*, 85(3):1249–1272, Feb 1999.

[129] James F. Ziegler, M.D. Ziegler, and J.P. Biersack. SRIM – The stopping and range of ions in matter (2010). *Nuclear Instruments and Methods in Physics Research Section B: Beam Interactions with Materials and Atoms*, 268(11-12):1818–1823, Jun 2010.

Index

Legendre polynomials, 48
Levy, 20
Lindhard-Sorensen correction, 128
linear attenuation coefficient, 182
linear attenuation coefficient , 184
linear energy transfer, 116
liquid drop model, 55
lognormal distribution function,
    87

Mann, 35
Marvin, 27, 44, 95
mass attenuation coefficient, 185
mass electronic stopping power,
    155
mass stopping power, 113, 116
matrix exponential, 21
matrix exponential function, 20
Maxwell distribution, 82
mean free path, 183
mean ionization potential, 132
mean lifetime, 3
medical physics, 177
Mimlitz, 49
Moller cross section, 154
Molybdenum 99, 18
Mott correction, 128
multichannel analyser, 80
multipole radiations, 44

narrow beam geometry, 182
neutrino, 25
Nickel, 6
Nishina, 205
Nitrogen-13, 33
normal distribution, 78
nuclear models, 55
Nuttall, 30

Pair production, 211
parity operator, 45

parity selection rule, 45
partial fractions, 11
Particle Data Group, 160
Pauli, 34
Pearson, 82
Penfold, 167
Perera, 49
permutations, 66
perturbation theory, 35, 115, 122,
    147, 190, 224
photoelectric cross section, 190
photoelectric effect, 188
photon attenuation, 182
Poisson distribution, 69, 79, 83
Polonium, 104
Polonium 210, 5
positron emission tomography, 53
Positron emission tomography,
    PET, 6
Positronium, 213
Potassium 40, 6
power law probability distribution,
    87
probability distribution, 61
Proton therapy, 142
PSTAR, 146, 170

Q-value, 21
quantum electrodynamics, 212
quantum mechanics, 30, 222
quantum tunnelling, 26

radiation dosimetry, 156
radiation length, 159
radiative stopping power, 160
radioactive decay law, 1, 181
radioactive equilibrium, 15
radiotherapy, 175, 190
Radium, 5
random error, 98

Printed in the United States
by Baker & Taylor Publisher Services